PLANET
EARTH

PLANET EARTH

HOW THE WORLD
WORKS, SHOWN IN
1000 PHOTOGRAPHS

**JOHN FARNDON, JACK CHALLONER,
ROBIN KERROD & RODNEY WALSHAW**

southwater

CONTENTS

INTRODUCTION.....6

PLANET EARTH.....8

ROCKS AND MINERALS.....68

VOLCANOES AND EARTHQUAKES.....128

WILD WEATHER.....189

INTRODUCING THE PLANET

Planet Earth is tiny compared to the vastness of Space. It is just one of nine planets circling the Sun. The Sun, in turn, is one of many billions of stars in one spiralling arm of a single galaxy. There are about 20 trillion other galaxies in the Universe. We are learning more about our planet in the context of space and about its own natural history. Such knowledge plays a crucial role in our lives, and a huge range of scientists is involved in studying the Earth and all its different aspects. Geologists study rocks and the Earth's history, while geomorphologists study landforms. Volcanologists study volcanoes and meteorologists study the weather. Ecologists study the way living things interact with their environment.

All if them are earth scientists. They very rarely work in a laboratory, because subjects such as glaciers, mountains and extreme weather systems need to be studied as or where they occur. Scientists often brave extreme conditions, sometimes risking their lives, to collect data and samples and test their theories in the field. They might measure the movement of a glacier or even a continent, map the lie of the land or

Volcanoes spew out molten rock from deep inside the earth. If we could predict erruptions, many lives would be saved.

work out the depth of chasms in an ocean floor.

Some of the most outstanding discoveries scientists have made in the last few decades are to do with the Earth's movement. You may think you're sitting still, not moving an inch, as you read this book. In fact, the world is whirling around as it spins on its axis, like a spinning top at over 800km/h – faster than any fairground ride. Simultaneously it is hurtling through the darkness of space as it journeys round the Sun at 80,000km/h! If a plane flew that fast, it would take less than 30 minutes to fly round the world. Even the land beneath our feet is in motion. The whole of the Earth's rigid outer shell is broken into fragments that are constantly shifting this way and that, jostling and crashing into each other. This process is called plate tectonics. It has helped explain how continents and oceans have changed through the ages, why earthquakes occur where they do, what makes volcanoes erupt, and even how mountain ranges are built up. The possibility that we can learn to predict when earthquakes might occur or volcanoes may erupt, and so save millions of lives, gives these investigations a special urgency.

Meteorologists have used remarkable advances in space and computer technology to radically enhance our ability to predict the weather. Using satellite eyes in space, they can monitor storms as

French Polynesia in the Pacific Ocean may become an island beneath the waves, if sea levels continue to rise.

they develop from little disturbances to monster cyclones. Computers allow meteorologists to crunch huge amounts of data about the atmosphere. The result is that we can forecast the weather three days in advance with a much greater precision than just ten years ago.

Some recent scientific discoveries are cause for concern. One is that certain gases that are released by processes such as making fast-food cartons or aerosol sprays, have attacked the layer of ozone gas high in the atmosphere. Without this protective ozone layer, we are exposed to some of the Sun's ultraviolet rays. These can cause cancer and blindness in humans and animals and damage crops. As yet, the ozone hole only appears over the North and South Poles in spring, but scientists calculate that it will gradually spread wider and last longer.

Waste gases from fuel, motor vehicles and factories build up in the atmosphere. This causes the atmosphere to warm as gases are trapped beneath the ozone layer. These gases make the atmosphere act like the glass panes of a greenhouse, trapping the Sun's heat. Previously the greenhouse effect kept the Earth comfortably warm but now the build up of these waste gases is forming a barrier that traps Earth's heat. Extra warmth makes the world a stormier, more violent place. It may turn grasslands into deserts and melt the polar ice caps, flooding huge areas. Scientists have found that even the vast Antarctic ice cap, which has been a chunk of solid ice for over 10 million years, is beginning to soften and melt in places. It is estimated that this will

The thick ice of a glacier is split as it moves down a valley, into pinnacles and deep crevasses caused by gravity and weight.

cause the sea level to rise at a rate of half a metre every 100 years, bringing floods to low-lying coastal regions. Some of these islands may even eventually become wholly submerged by the oceans.

PLANET EARTH

*The study of the Earth is quite new.
Not long ago, most people thought that the Earth
was just a few thousand years old.
Then less than 200 years ago,
scientists began to realize that the Earth
is immensely old, well over four thousand million
years, and it has been through an
incredible series of changes in its life.
The more we learn about these changes
and how they affect us, the more amazing
we realize the Earth's mysteries are.*

Author: John Farndon
Consultant: Rodney Walshaw

THE UNIQUE PLANET

The Earth is the third planet out from the Sun, the third of the nine planets that make up the Solar System. It is a big round ball of rock with a metal core, wrapped around in a thin blanket of colourless gases called the atmosphere. From a distance, the Earth shimmers like a blue jewel in the darkness of space, because more than 70 per cent of its surface is covered with the oceans. No other planet has this much water on its surface. Jupiter's moon Europa is the only other place with much water, but it is so far away from the Sun that the water is frozen. Earth is neither so near the Sun that water is turned to steam, nor so far that it freezes. Only on the very ends of the Earth, at the poles, is water permanently frozen in ice caps. In places, rock sticks up above the ocean waters to form half a dozen large continents and thousands of smaller islands.

Round Earth

In the last 40 years, spacecraft have been able to take photographs of the Earth from space, so we can see that it is round. The Ancient Greeks suspected that the Earth was round 2,500 years ago, because they saw how ships gradually disappeared over the horizon. But for 2,000 years, many people continued to believe that it was flat. Only when the ship of explorer Ferdinand Magellan sailed round the Earth in 1522 were people finally convinced.

Water world

An island in the Indian Ocean demontrates all that is unique about planet Earth. The combination of vegetable and plant life, land and sea is found only on Earth – it exists nowhere else in the known universe. Life on Earth exists because it is a watery planet. Water is still, however, a precious resource. Almost 97 per cent is salt water in the oceans, and three-quarters of the remaining fresh water is frozen.

The Solar System

Earth's nearest companions in space are the planets Venus and Mars, all of which are much the same size. Tiny Mercury nearest the Sun, and even tinier Pluto farthest away, are the terrestrial (rocky) planets of the Solar System. The other four planets – Jupiter, Saturn, Uranus and Neptune – are gigantic by comparison and are made not of rock, but gas. It was once thought the Solar System was unique in the Universe, but astronomers have spotted planets circling distant stars. So there may be another Earth out there after all.

Mercury

Venus

Earth

Mars

The Sun

Man on the Moon

On 20 July 1969, the American astronaut Neil Armstrong stepped from the landing module of the *Apollo 11* spacecraft on to the Moon's surface. It was the first time that a human had ever set foot on another world. It was an exciting experience, and it reminded the astronaut just how special the Earth is. The Moon is really very near to us in space, yet it is completely lifeless, a desert of rock with no atmosphere to protect it from the Sun's dangerous rays and no water to sustain life.

The surface of Mars

Mars is like Earth in size, so people hoped it might have life of its own. Astronomers were once convinced that marks on the surface were canals built by Martians. Sadly, space probes to Mars have found no signs of life. As this photo from the *Pathfinder* mission shows, it is nothing but rocks and dust. But a few years ago NASA scientists found what could be a fossil of a microscopic organism in a rock that fell from Mars.

Jupiter

Saturn

Uranus

Neptune

Pluto

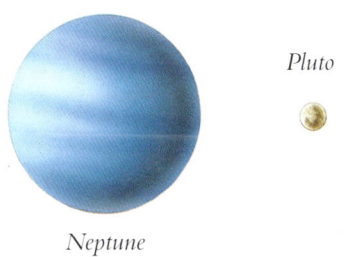

SPINNING PLANET

You will need: *felt-tipped pen, plastic ball, piece of thin string, modelling clay, torch.*

NIGHT AND DAY

1 Draw, or cut out and glue, a shape on the ball to represent the country you live in. Stick the string to the ball with modelling clay. Tie the string to a rail, such as a towel rail, so that the ball hangs freely.

T HE EARTH is like a giant ball spinning in the darkness of space. The only light falling on it is the light of the Sun glowing 150 million kilometres away. As it spins, the Earth also moves around the Sun. The two ways of moving explain why night and day, and the seasons, occur. At any one time half the world is facing the Sun and is brightly lit, while the other half is facing away and is in darkness. As the Earth spins on its axis, the dark and sunlit halves move around, bringing day and night to different parts of the world.

The Earth is always tilted in the same direction. So when the Earth is on one side of the Sun, the Northern Hemisphere (the world north of the Equator) is tilted towards the Sun, bringing summer. At this time, the Southern Hemisphere (the world south of the Equator) is tilted away, bringing winter. When the Earth is on the other side of the Sun, the Northern Hemisphere is tilted away, bringing winter, while the south is in summertime. In between, as the Earth moves around to the other side of the Sun, neither hemisphere is tilted more than the other one towards the Sun. This is when spring and autumn occur. These two experiments show how this happens. The ball represents the Earth and the torch is the Sun.

2 Shine the torch on the ball. If your country is on the half of the ball in shadow on the far side, then it is night because it is facing away from the Sun.

3 Your home country may be on the half of the ball lit by the torch instead. If so, it must be daytime here because it is facing the Sun. Keep the torch level, aimed at the middle.

4 Turn the ball from left to right. As you turn the ball, the light and dark halves move around. You can see how the Sun comes up and goes down as the Earth turns.

THE SEASONS

You will need: felt-tipped pen, plastic ball, bowl just big enough for the ball to sit on, torch, books or a box to set the torch on.

1 Use the felt-tipped pen to draw a line around the middle of the ball. This represents the Equator. Sit the ball on top of the bowl so that the Equator line is sloping gently.

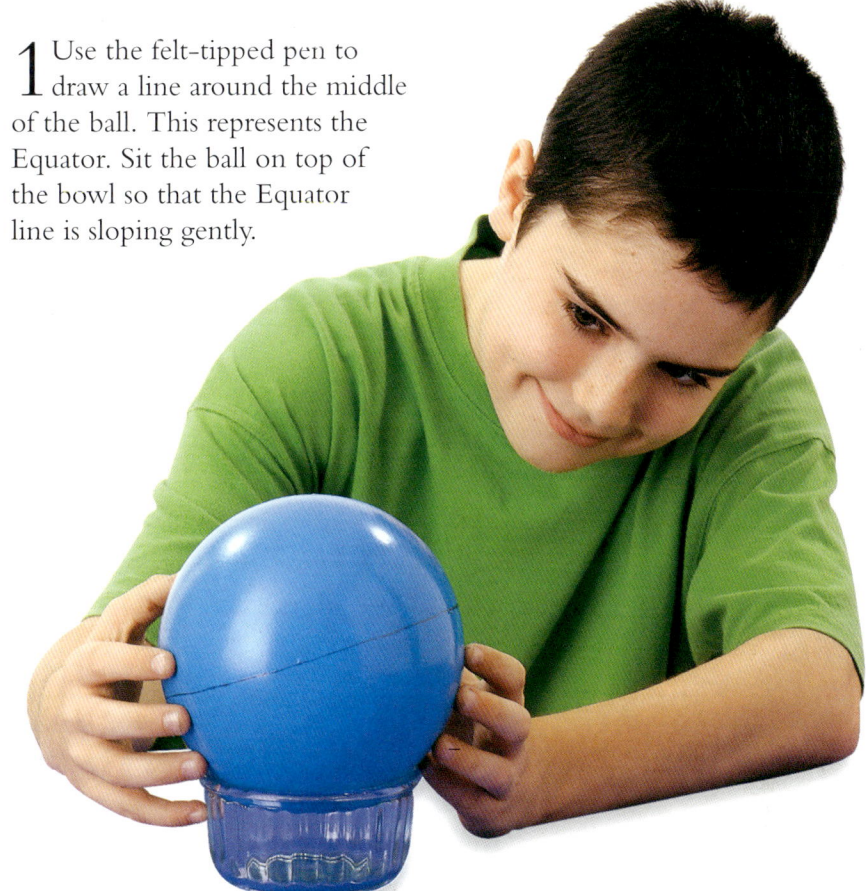

2 Put the torch on the books so it shines just above the Equator. It is summertime on the half of the ball above the Equator where the torch is shining, and winter in the other half of the world.

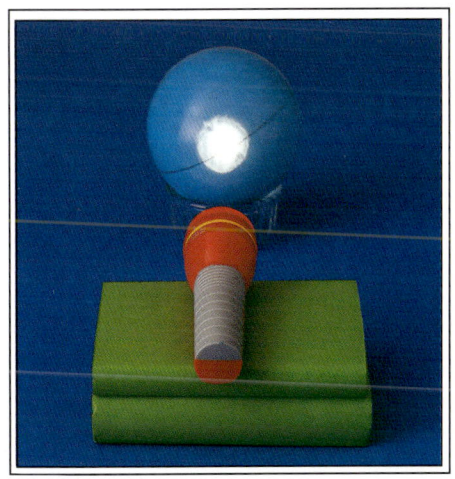

3 Shine the torch on the Equator, as shown. It sheds an equal amount of light in each hemisphere. This is the equivalent of spring and autumn, when days and nights are of similar length throughout the world.

Half Moon

The Moon shines not because it gives out light itself but because it reflects the light of the Sun. Just as half of the Earth is always lit by the Sun, so is half of the Moon. The Moon appears to change shape during the month, from a crescent to a disc and back again. This is because it moves round the Earth and so we see its sunlit side from different angles. When we see a full moon, we are seeing all of the sunlit side. The Sun and stars are our only sources of light from space.

EARTH STORY

THE SOLAR System formed 4,570 million years ago from debris left over from the explosion of a giant star. As the star debris spun around the newly formed Sun, it began to congeal into balls of dust. Quickly, the dust clumped into tiny balls of rock called planetesimals, and the planetesimals clumped together to form planets such as the Earth and Mars. At first, the Earth was little more than a ball of molten rock. Then, when the Earth was about 50 million years old, it is thought that a giant rock cannoned into it with such force that the rock melted. The melted rock cooled to become the Moon. The Earth itself was changed forever. The shock of the impact made iron and nickel collapse to Earth's centre, forming a core so dense that atoms fuse in nuclear reactions. These reactions have kept the Earth's centre ferociously hot ever since. The molten rock formed a thick mantle around the core, kept churning slowly by the heat.

A volcanic landscape

It is hard to imagine what the Earth's surface was like in the very early days, but you could get a good idea by staring down into the mouth of an active volcano. The collision of rocks that actually created the Earth left it incredibly hot – hot enough to melt the rock it is made from. The whole planet was just one giant, seething red hot ball of magma (molten rock).

1

2

3

4

The formation of planet Earth

The Solar System and planet Earth formed when gravity began to pull the debris left over from a giant star explosion into clumps (1). At first, Earth was just a molten ball, and for half a billion years it was bombarded by meteorites (2, 3). By 3.8 billion years ago, things were calming down (4, 5). The crust and atmosphere had formed, and very solid lumps of rock on the planet's surface were forming the first continents (6).

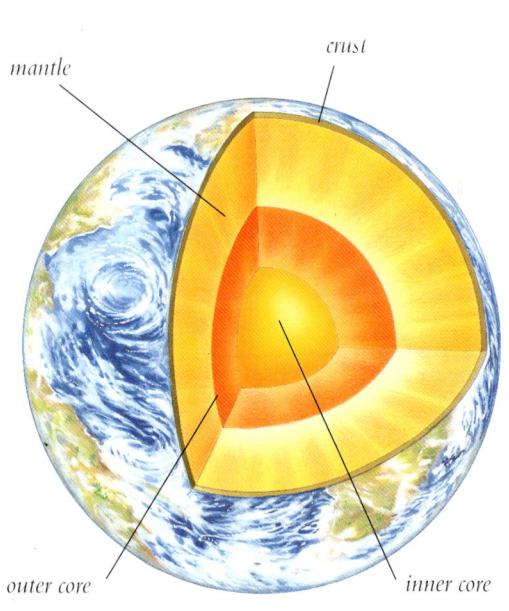

mantle

crust

outer core

inner core

The layers of the Earth

The Earth is not a solid ball of rock. The collision with the giant rock that formed the Moon caused the Earth to separate into several distinct layers. On the outside is a thin layer of solid rock, up to 40km thick, called the crust. Below the crust is a thick layer of soft, semi-molten rock known as the mantle. At the centre is a core of iron and nickel. The outer core is so hot it is molten, but the pressure in the inner core is so intense it cannot melt, even though temperatures here reach 3,700°C.

5

6

Steam power

These hot springs in Iceland are sending out jets of steam heated by the Earth's interior. But they are tiny compared to the huge amounts of steam and gas that must have billowed out from the hot Earth in its early days. Two hundred million years after the birth of planet Earth, volcanic fumes formed the atmosphere. All that was missing was oxygen – the vital ingredient added later by plants.

Auroras

The spinning of the Earth swirls the iron in its core, turning it into a huge electric dynamo – and the electricity turns the Earth into a magnet. The effect of Earth's magnetism is felt tens of thousands of kilometres out in space. Indeed, there is a giant cocoon of magnetism around the Earth called the magnetosphere. This shields us from harmful electrically charged particles streaming from the Sun. There are small gaps in this shield above the poles. Here charged particles collide with the atmosphere and light up the sky in spectacular displays of light called auroras.

MAGNETIC EARTH

The sailor's guide
Until the age of satellites, the magnetic compass was the sailor's main tool for finding the way at sea. It had to work in any weather, at any time of day, and on a rolling ship.

MAKE A COMPASS

You will need: bar magnet, steel needle, slice of cork, sticky tape, small bowl, jug of water.

THE EARTH behaves as if there is a giant iron bar magnet running through its middle from pole to pole. This affects every magnetic material that comes within reach. If you hold a magnet so that it can rotate freely, it always ends up pointing the same way, with one end pointing to the Earth's North Pole and the other to the South Pole. This is how a compass works – the needle automatically swings to the North. These projects show you how to make a compass, and how you can use it to plot a magnetic field like the Earth's. The Earth's magnetic field is slightly tilted, so compasses do not swing to the Earth's true North Pole, but to a point that is a little way off northern Canada. This direction is known as magnetic north.

The Earth's magnetism comes from its inner core of iron and nickel. Because the outer core of the Earth is liquid and the inner core solid, the two layers rotate at different rates. This sets up circulating currents and turns it into a giant dynamo of electrical energy.

1 To turn the needle into a magnet, stroke the end of the magnet slowly along it. Repeat this in the same direction for about 45 seconds. This magnetizes the needle.

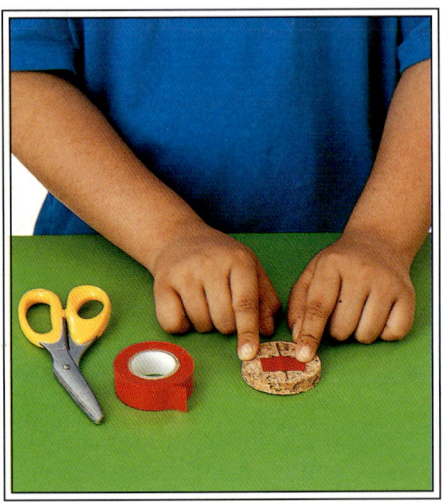

2 Place the magnetized needle on the piece of cork. Make sure that it is exactly in the middle, otherwise it will not spin evenly. Tape the needle into position.

3 Fill the bowl near to the brim with water and float the cork in it. Make sure the cork is exactly in the middle and turns without rubbing on the edges of the bowl.

4 The Earth's magnetic field should now swivel the needle on the cork. One end of the needle will always point to the north. That end is its north pole.

MAGNETIC FIELD

You will need: *large sheet of paper, bar magnet, your needle compass from the first project, pencil.*

1 Lay a large sheet of paper on a table. Put the magnet in the middle of the paper. Set up your needle compass a few centimetres away from one end of the magnet.

2 Wait as the compass needle settles in a particular direction as it is swivelled by the magnet. Make a pencil mark on the paper to show which way it is pointing.

3 Move the compass a little way towards the other end of the magnet. Mark a line on the paper to show which way the needle is pointing now.

4 Repeat Step 3 for about 25 different positions around the magnet. Try the compass both near the magnet and further away. You should now have a pattern of marks.

Magnetic protection

The effects of Earth's magnetism extend 60,000km out into space. In fact, there is a vast magnetic force field around the Earth called the magnetosphere. This traps electrically charged particles and so protects the earth from the solar wind – the deadly stream of charged particles whizzing from the Sun.

5 Look at the pattern of marks you have made on the paper. They should form a series of rings around the magnet, like layers of an onion. Earth's magnetic field is shaped like this.

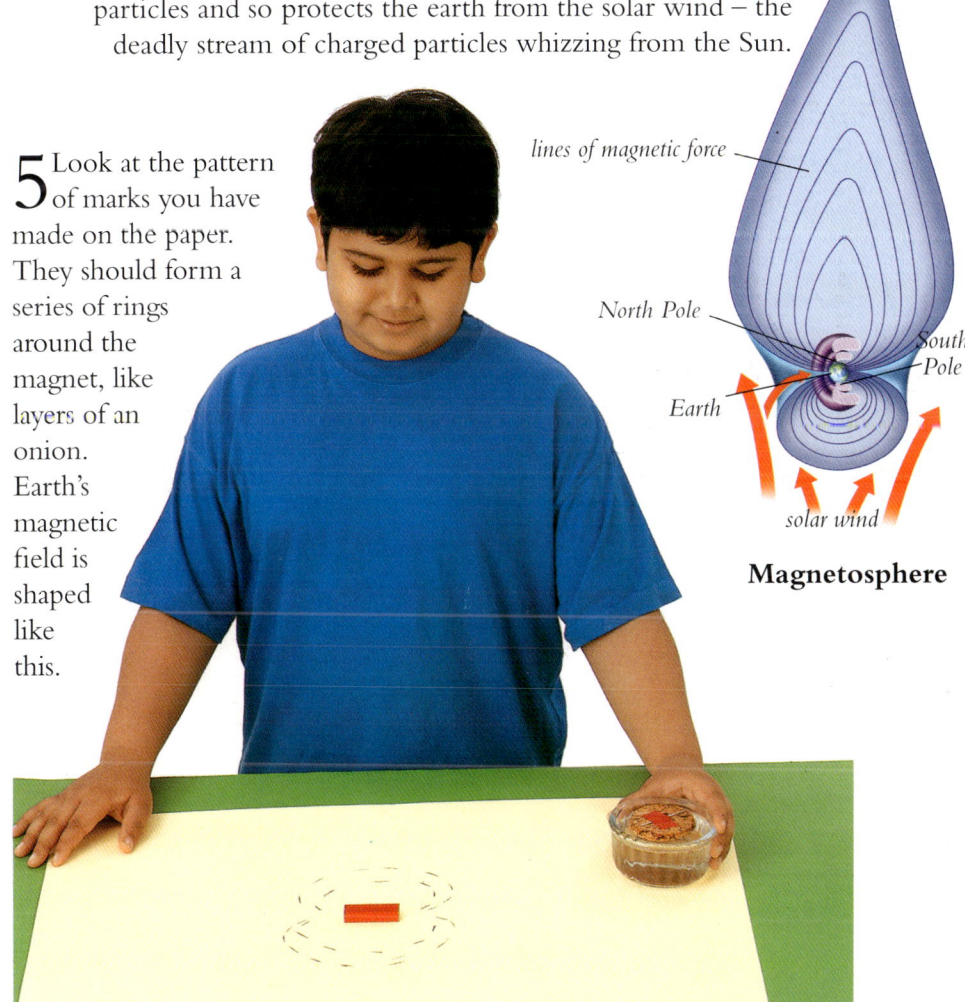

tail of magnetosphere blown out by solar wind

lines of magnetic force

North Pole

South Pole

Earth

solar wind

Magnetosphere

LIFE-SUPPORT SYSTEM

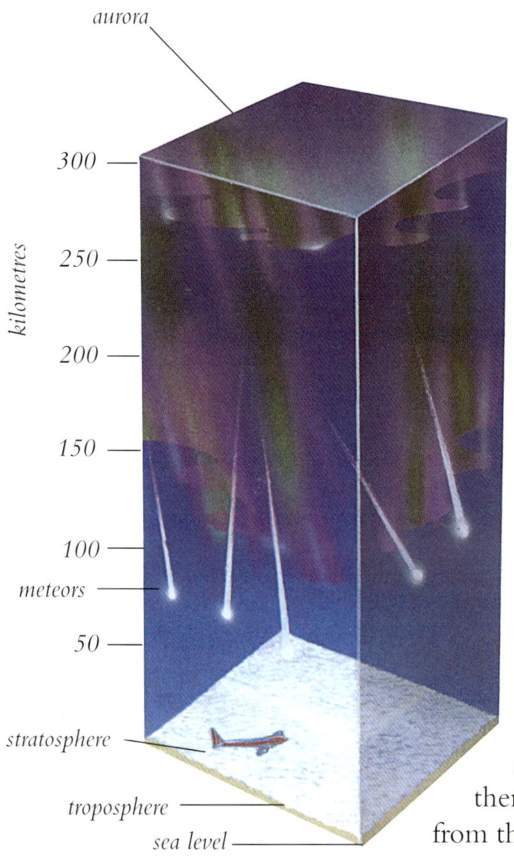
aurora
300
250
200
150
100
meteors
50
kilometres
stratosphere
troposphere
sea level

LIFE ON Earth exists only because of the atmosphere. This thin blanket of gases, tiny water drops and dust is barely thicker on the Earth than the skin is on an orange. At its thickest, it is less than 1000km deep. The atmosphere has no colour or taste, yet it is much more interesting than it may appear, because it is a surprisingly complex mixture of gases. Over 99 per cent of the atmosphere is made up of two gases, nitrogen (78 per cent) and oxygen (21 per cent). It also contains argon, carbon dioxide, water vapour and minute traces of many other gases, such as helium and ozone.

Without the atmosphere, planet Earth would be as lifeless as the Moon. The atmosphere gives us air to breathe and water to drink, and keeps us warm. The atmosphere also acts as a shield, filtering out the Sun's harmful rays and protecting us from falling meteorites.

Layers of air
We live in the densest, bottom layer of the atmosphere, called the troposphere. This is where most of the water vapour is, and all the weather. The calmer stratosphere is 12km-50km above ground, and gets warmer higher up. The mesosphere above is thinner in gases and very cold. Beyond is the thermosphere, heated by ultraviolet rays from the Sun to a temperature of 2,000°C.

FACT BOX

- If all the water vapour in the air suddenly condensed, it would make the oceans 3cm deeper.
- Over a billion tonnes of salt join the atmosphere from sea spray each year, and over a quarter of a billion tonnes of soil dust.
- The stratosphere glows faintly at night. This airglow is created by sodium or salt spray from the sea that is heated by the sun during the day.
- Early radio signals were bounced round the world off electrically charged particles in the thermosphere.

A thin veil
Earth's atmosphere can be seen in this photograph taken from the Space Shuttle. The Sun is beyond the horizon, just catching the edge of the new Moon high above. As the Sun's rays are scattered this way and that, through the dense, watery, dusty lower layers of the atmosphere, they turn red and orange. Further up, where the air is thinner and there is no dust or water, the atmosphere is blue. This is what you see when you look up on a clear day. The sky is blue because gases in the air reflect mostly blue light from the sun. Dust and moisture scatter other colours and dilute the blue.

Stratocruiser

Aeroplanes climb high up above 12km as soon as they take off. By doing this, they break out of the troposphere, where they would be buffeted by the weather, and soar into the calm stratosphere, where there is no weather at all. At this height, the passengers need to be inside a pressurized cabin. Our bodies are built to cope with the pressure of air at ground level. Up in the stratosphere where the air is thin, the pressure is too low.

Outer limits

The Space Shuttle orbits Earth in space about 300km up. This is the outermost layer of Earth's atmosphere, the exosphere. Here the gases are rarefied, that is, they are very few and far between. Above the exosphere, the atmosphere fades away into empty space. There is no oxygen at this height – just nitrogen and more of the lighter gases such as hydrogen.

Thin air

Gravity pulls most of the gases in the air into the lowest layers of the atmosphere. Seventy-five per cent of the weight of the gases in the air is squashed into the troposphere, the lowest one per cent of the atmosphere. The air gets thinner very quickly as you climb. Mountaineers climbing the world's highest peaks need oxygen masks to breathe because there is much less oxygen at this height.

Making clouds

Clouds are made from billions of tiny droplets of water and ice crystals so tiny that they float on the air. The droplets are formed from water vapour (a bit like steam) rising from the sea as it is warmed by the Sun. Air gets steadily colder higher up, and as the vapour rises it cools down. Eventually the water vapour gets so cold that it turns into tiny droplets of water and forms a cloud.

STIRRINGS OF LIFE

LIFE PROBABLY began on Earth entirely by chance, about 3.8 billion years ago. The early Earth was a hostile place. It seethed with erupting volcanoes, was washed by oceans of warm acid and enveloped in toxic fumes. But these could have been just the right conditions to start life. Amino acids, the building blocks from which the first living cells were formed, may have been created by chance as small molecules were fused together by lightning bolts that surged through the stormy air.

Recently, amazing microscopic bacteria called archaebacteria were found living on black smokers on the ocean floor, in conditions as hostile as the early Earth. It may be that bacteria such as these were the first living cells, feeding on the chemicals spewed out by volcanoes. Another kind of bacteria, called cyanobacteria or blue-green-algae, appeared later.

Geological time

Much of what we know about the story of life on Earth comes from fossils, which are the remains of organisms in rocks. Layers of rock form one on top of the other, so the oldest is usually at the bottom, unless they have been disturbed. By studying rocks from each era, palaeontologists (scientists who study fossils) have slowly built up a picture of how life has developed over millions of years.

Ancient lifeform

Archaebacteria are the oldest known form of life. They seem to be able to thrive in the kind of extreme conditions that would kill almost everything else, including other bacteria. Some have been found living in boiling sulphur fumes on volcanic vents on the sea floor. This one was found in Ace Lake, Antarctica, and can survive in incredibly cold temperatures, living off carbon dioxide and hydrogen.

Hardy organisms

One of the most remarkable discoveries of recent years has been tall volcanic chimneys on the ocean floor that belch hot black smoke. These black smokers, or hydrothermal vents, are home to a community of amazing organisms that actually thrive in the scalding waters and toxic chemicals that would kill other creatures. These communities give important clues to how life could have started in the similarly hostile conditions on the primeval Earth four billion years ago.

First lifeforms (bacteria) appear, and give the air oxygen.

No life on land, but shellfish flourish in the oceans.

Early fish-like vertebrates appear. The Sahara is glaciated.

First land plants. Fish with jaws and freshwater fish.

First insects and amphibians. Ferns and mosses as big as trees.

Vast warm swamps of fern forests which form coal. First reptiles.

Precambrian time

Cambrian Period 590 million years ago

Ordovician Period 505 million years ago

Silurian Period 438 million years ago

Devonian Period 408 million years ago

Carboniferous 360 million

Puffs of oxygen

Early archaebacteria left no trace. The oldest signs of life are microscopic threads in rocks dating back 3.5 billion years. The threads are like the blue-green algae called cyanobacteria that live in the oceans today. The tiny algae changed the world by using sunlight to break carbon dioxide in the air into carbon and oxygen. They fed on the carbon and expelled oxygen. The little puffs of oxygen seeped into the air, filling it with oxygen and preparing the way for life as we know it.

Ancient slime

The oldest proofs of life are stony mounds called stromatolites. Some of those at Fig Tree Rock in South Africa, date back 3.5 billion years. Stromatolites are the fossilized remains of huge colonies of slimy bacteria with a thin layer of cyanobacteria on top. The cyanobacteria obtain their food from sunlight, while the bacteria feed on dead cyanobacteria. Stromatolites called conyphytons grew up to 100m high.

Clouds of life

Most scientists believe that the chemicals of life were assembled on Earth. Some, such as the late Fred Hoyle, believe life came from space. Clouds of stardust in space, called giant molecular clouds, do contain basic life chemicals. These include huge amounts of ethyl alcohol – the alcohol in drinks.

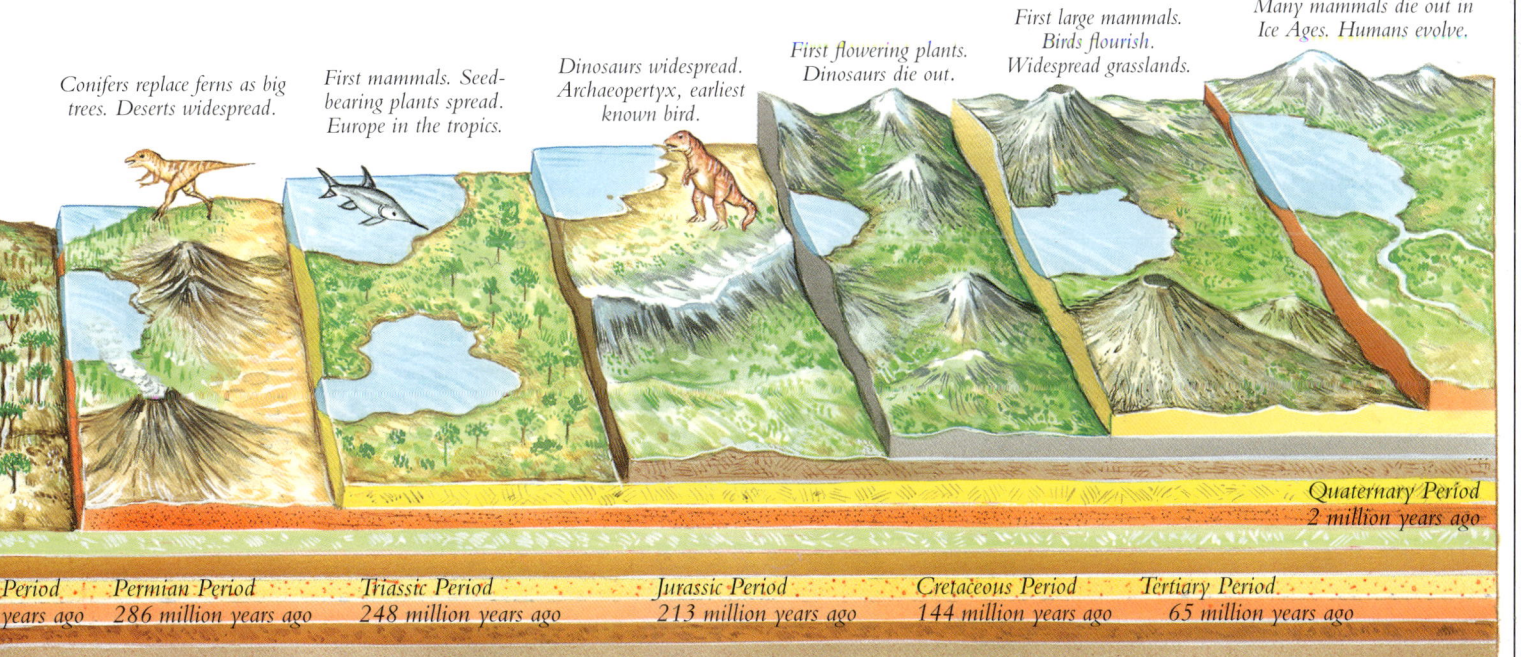

Conifers replace ferns as big trees. Deserts widespread.

First mammals. Seed-bearing plants spread. Europe in the tropics.

Dinosaurs widespread. Archaeopertyx, earliest known bird.

First flowering plants. Dinosaurs die out.

First large mammals. Birds flourish. Widespread grasslands.

Many mammals die out in Ice Ages. Humans evolve.

Quaternary Period 2 million years ago

Period years ago	Permian Period 286 million years ago	Triassic Period 248 million years ago	Jurassic Period 213 million years ago	Cretaceous Period 144 million years ago	Tertiary Period 65 million years ago

EVOLUTION OF SPECIES

FOR THE first three billion years of Earth's history, the only life was in the form of microscopic, single-celled organisms in the oceans. Then about 700 million years ago, the first real animals appeared. These were creatures such as jellyfish and sponges made from many kinds of cells, each one suited to a different task. These creatures were soft and left few traces. Over the next 100 million years, animals with shells and bones appeared. Hard parts fossilize easily, and from this time – the start of the Cambrian Period around 590 million years ago – there are very many fossils. From these, scientists have pieced together the story of how different species have come and gone. These include not only the dinosaurs, the gigantic reptiles that dominated the planet for 155 million years, but the first human-like creatures.

Ammonites

Most fossils that are found are shellfish, because they are so easily preserved. One of the commonest shellfish fossils is the ammonite, a creature a bit like a squid that lived inside a spiral shell. Ammonites lived in the oceans around the time when dinosaurs dominated the land – from 220 to 65 million years ago. They are very common and evolved quickly into different types, that can be slotted into specific periods of time. This means that geologists can use them to date the rocks they are found in.

First plants

Life began in the oceans and moved on to land only gradually from around 400 million years ago. The first trees were not like today's trees. They were the ancestors of today's ferns and cycads, like this one in Australia. Unlike modern trees, which grow from seeds and flowers, ferns and cycads grow from tiny spores. Today, ferns and cycads are usually quite small, but around 300 million years ago, they grew into huge, tall forests.

Living fossils

Coelacanths are remarkable fish that first appeared 400 million years ago. They were once thought to have died out 65 million years ago, since no more recent fossils have been found. But in 1938 a living coelacanth was found in the Indian Ocean. The fish have muscular, limb-like fins, and it is from species like these that the first land creatures evolved. The limb-like fins developed into legs so that the fish could haul themselves across mud flats.

- The first creatures with bones appeared 570 million years ago.

- The first plants and insects appeared on land 400 million years ago.

- The first amphibians crawled on to the land 350 million years ago.

- The biggest carnivore, the sea reptile lipluridon, lived 150 million years ago and grew to 25m long.

- The biggest land creature, the dinosaur Brachiosaurus, grew to 25m long.

- Evolution was not a smooth, continuous process but was broken by periodic mass extinctions of species.

Lifestyle

Dinosaurs dominated Earth from 225 to 65 million years ago, and then they mysteriously died out altogether. As palaeontologists have found more and more dinosaur fossils, they have been able to work out in detail what they looked like and how they lived. Stegosaurs were plant-eating dinosaurs that lived from about 180 to 80 million years ago. They had two rows of pointed plates down their back and a spiny tail that they swung to protect themselves from predators.

Frozen baby

Woolly mammoths were hairy, elephant-like creatures that lived in northern Asia and North America until about 10,000 years ago. Mammoths were probably hunted to extinction by early humans, but bodies have been found frozen in the permafrost (permanently frozen ground) of Siberia. They are so well preserved that some Japanese scientists hope to recreate them. The idea is to take from the bodies DNA, the chemical that carries their genetic code. The DNA could be used to create an embryo, which could be implanted in a living elephant.

Digging for life

When miners dig coal out of a mine, they are digging out the history of life on Earth. Coal is the fossilized remains of forests of giant club mosses and tree ferns that grew in vast tropical swamps some 350 million years ago, in the Carboniferous Period. Over millions of years, layer upon layer of dead plants sank into the swamp mud. As they were buried deeper and deeper, they were squeezed dry and became hard, slowly turning to almost pure carbon. The deeper the remains were buried, the more completely they have turned to carbon. So, coal near the surface is brown. Deep down it is jet black.

COMPETITIVE GENES

You will need: *medium-size bowl, spoon, 100g sea salt, liquid fish food, brine shrimp eggs from an aquarium shop, magnifying glass.*

COMPETING FOR LIFE

TODAY, THERE is an astonishing variety of life on Earth, with millions of species of animals and plants. Yet each one has its own natural home or way of life. Every living thing is adapted (suited) to its surroundings. In 1859, the English naturalist Charles Darwin explained all this with his theory of evolution by natural selection. This theory shows how over millions of years species gradually change or evolve. As they change, they adapt to their surroundings, and new species emerge. Evolution like this depends on the fact that no two living things are alike. So some may start life with features that make them better able to survive, as the first experiment shows. An animal, say, might have long legs to help it escape from predators. Individuals with such valuable features have a better chance of surviving. They may also have offspring that inherit these features, as in the way shown in the Strong Genes experiment. Slowly, over generations, better adapted animals and plants flourish while others die out or find a new home. In this way, species gradually evolved.

Non-survivor
The ancient stingray fossilized here survived for millions of years. Then, however, this ancient species suddenly became extinct. Conditions, such as the climate, changed, and the fish did not adapt to the new conditions fast enough to survive.

1 To see how some eggs survive and others don't, pour 1 litre of warm water into a bowl. Stir in the sea salt until it dissolves. When the water is cool, add a few drops of fish food.

2 Sprinkle in a spoonful of shrimp eggs. Leave the bowl in a warm place such as an airing cupboard so that the water will stay at 21°C.

3 After a few days, some shrimps will hatch into larvae. Stir the water once a day and scoop out a spoonful. Be careful not to disturb the larvae too much.

4 Using a magnifier, count how many eggs and larvae you see on the spoon. When adults appear, count these too. Only the strongest will survive and grow into adults.

The lottery of life

Only a tiny proportion of this dragonfly's thousands of eggs will live to adulthood. Dragonflies generally have adapted well to changing conditions. They are the longest-surviving of all insect species. More than 300 million years ago, they were the first animals to fly.

STRONG GENES

You will need: pack of ordinary playing cards. This is a game for three or more players.

1 Deal seven cards to each player. The cards are your supply of genes for life. The suit of diamonds represents strong genes that give you a better a chance of survival.

2 Each player lays down a card in turn, following suit. If you cannot follow suit, play a diamond – a strong gene. If you play a diamond, always pick it up and save for the next deal.

3 The player who plays the highest card (or a diamond) wins the round, and begins the next. Once all the cards have been played, the player who has won the fewest rounds drops out.

4 Deal six cards to each survivor. Each player then picks up their diamonds from the previous round. Discard one card from the six for each diamond you pick up.

5 Play out all the cards in rounds as before. Again the player who wins the fewest rounds must drop out. Repeat Step 4 dealing five cards each. Play again with four cards, then three, then two, then one. The last player survives life's game.

BROKEN EARTH

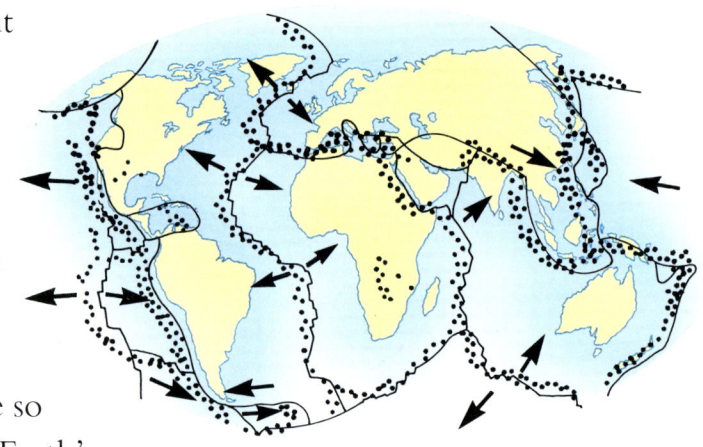

THE SURFACE of the Earth is not in one piece, but cracked like a broken eggshell, into 20 or so giant slabs. The giant slabs, called tectonic plates, are huge, thin pieces of rock thousands of kilometres across but often little more than a dozen kilometres thick. The plates are not fixed in one place, but are slipping and sliding around the Earth all the time. They move very slowly – at about the pace of a fingernail growing – but they are so gigantic their movement has dramatic effects on the Earth's surface. The movement of the plates causes earthquakes, pushes up volcanoes and mountains, and makes the continents move. Once, the continents were all joined together in one huge continent that geologists call Pangea, surrounded by a giant ocean called Panthalassa. Around 200 million years ago the plates beneath Pangea began to split up and move apart, carrying fragments of the continent with them. These fragments slowly drifted to the positions they are in now.

Surface sections
The Earth's rigid shell is called the lithosphere, from the Greek word *lithos* (stone). It is broken into the huge fragments shown on this map. The African plate is gigantic, underlying not only Africa but half of the Atlantic Ocean too. The Cocos plate under the West Indies is quite small. Black dots mark the origins of major earthquakes over a year. Note how they coincide with the plate margins.

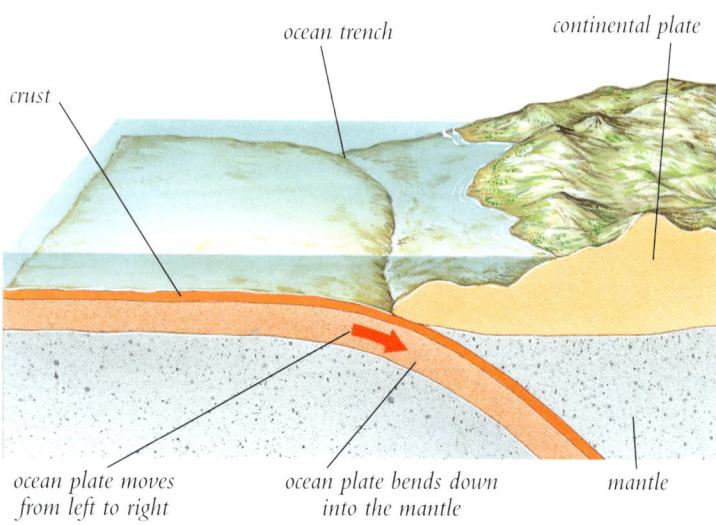

crust

ocean trench

continental plate

ocean plate moves from left to right

ocean plate bends down into the mantle

mantle

Plates in collision
In many places, tectonic plates are slowly crunching together with enormous force. As they collide, one plate may be forced under the other in a process called subduction, which means drawing under. The plate is completely destroyed as the plate slides down into the heat of the Earth's mantle. Earthquakes are often generated as it slides down, and the melting rock bubbles up as violent volcanoes. Subduction zones occur right around the western edge of the Pacific Ocean.

Lowest place on Earth
When one plate is forced down beneath another, it can open up deep trenches in the ocean floor. These trenches are the lowest places on the Earth's surface. They plunge so far down that the water is darker than the blackest night. These dark abysses have been never been fully explored, but this photograph was taken at the bottom of the world's deepest trench, the Marianas Trench, at a depth of over 10,000m.

Two-way friction

In some places, the plates are neither crunching together nor pulling apart. Instead, they are juddering sideways past each other. This is happening at the San Andreas fault in California. The giant Pacific plate is sliding at a rate of 6cm a year north-west past the North American plate. This has set off earthquakes that have rocked the cities of San Francisco and Los Angeles.

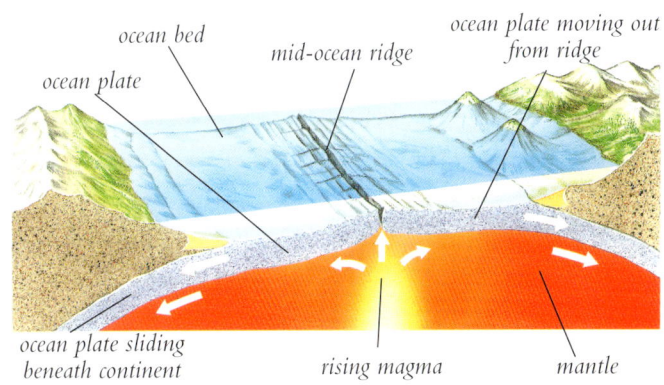

ocean bed   mid-ocean ridge   ocean plate moving out from ridge

ocean plate

ocean plate sliding beneath continent   rising magma   mantle

Mid-ocean ridge

Right down the middle of the sea bed in the Atlantic Ocean, there is a giant crack where the tectonic plates are pulling apart. Here, molten rock from the Earth's interior wells up into the crack and freezes on to the edges of the plates. This creates a series of jagged ridges called the mid-ocean ridge. As the plates move apart and new rock is added on the edges, so the sea floor spreads wider.

A string of islands

As one colliding plate is forced down beneath another into the Earth's hot interior, it melts. The melting rock is squeezed into cracks in the overlying plate and bursts on to the surface as a line of volcanoes. If this line is in the sea, the volcanoes form a long, curving string of islands called an island arc. Many of the islands in the western Pacific formed in this way.

FACT BOX

• The west coast of Africa looks as if it would slot into the east coast of South America because they were once joined together.

• New York is moving 2-3cm farther from London every year.

BUILDING MOUNTAINS

A FEW of the world's highest mountains, such as Mount Kilimanjaro in Africa, are lone volcanoes, built up by successive eruptions. Most high mountains are part of great ranges that stretch for hundreds of kilometres. Mountains look as if they have been there forever, but geologically they are quite young. They have all been thrown up in the last couple of hundred million years – the last 25th of the world's history – by the huge power of the Earth's crust as it moves. The biggest ranges, such as the Himalayas and the Andes, are fold mountains. These are great piles of crumpled rock pushed up by the collision of two of the great plates that make up the Earth's surface. Folding opens up many cracks in the rock and the weather attacks them, etching the mountains into jagged peaks and knife-edge ridges. Some mountain ranges, in the central parts of plates, are huge blocks of the Earth's crust that have been pushed up as the plate was stretched.

Lone volcano

Mount Kilimanjaro, on the border of Kenya and Tanzania, is Africa's highest mountain, reaching 5,895m. This distinctive volcanic cone rises in isolated majesty from the surrounding grassland. It was formed where a rising plume of hot rock in the Earth's interior burned its way up through the tectonic plate that underlies East Africa. Mount Kilimanjaro is high enough to be snowcapped, even though it lies near the Equator.

Bow-wave

Geologists used to think that rock crumples to form fold mountains, such as the Alps, in rigid layers. Now they are beginning to think that because this all happens so slowly, the rock flows almost like very thick treacle. They suggest that as one tectonic plate collides with another, the rock is pushed up like the v-shaped bow-wave in front of a boat. Like a very slow bow-wave, the mountains are continually piling up in front of the plate and flowing away at the side.

Continental crunch

The incredible contortions created by rock folding even on a small scale are clearly visible in the zigzag layers of shale. These rocks in Devon, England were folded about 250 million years ago. This was when a mighty collision between two continents that formed the single giant continent of Pangea. Present-day Devon was one of the places that was squeezed in the gap.

Waves in the Andes

Using the latest satellite techniques, geologists have surveyed many of the world's high mountain ranges, including the Andes, the Himalayas, the European Alps and New Zealand. What they have found is something remarkable. When compared to surveys from a century ago, mountain peaks in these ranges have moved exactly as if they are flowing very slowly. So the folds in the rock you see in this photograph may not be folds but very, very slow and stiff waves.

How faults occur

The slow, unstoppable movement of tectonic plates puts rock under such huge stress that it sometimes cracks. Such cracks are called faults. Where they occur, huge blocks of rock slip up and down past each other, creating cliffs. In places a whole series of giant blocks may be thrown up together, creating a new mountain range. The Black Mountains in Germany are an example of block mountains formed in this way.

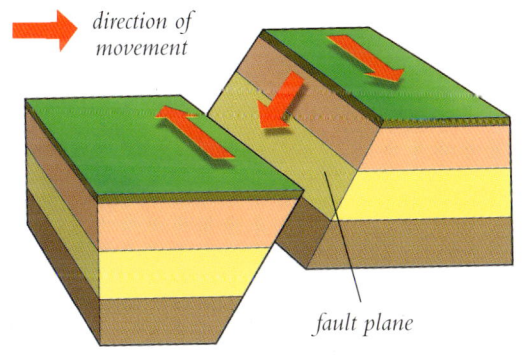

direction of movement

fault plane

The Earth splits open

The Great Rift Valley runs for more than 3,000km down the east of Africa from the Red Sea. The movement of the Earth's tectonic plates not only builds up mountains, it also opens up great valleys. Where plates are pulling apart, or magma is pushing up underneath, the land can split like the crust of a pie in the oven. Land can drop along this crack to form a rift valley.

Zagros Mountains, Iran

For the last 100 million years, the tectonic plates carrrying Africa and Arabia have been ploughing northwards into Eurasia. The tremedous impact has crumpled the edge of Eurasia and thrown up a belt of mountains stretching all the way from southern Europe, through the Aegean and Turkey to the Zagros Mountains of Iran. The Turkish and Iranian end remains very active, building up the mountains still higher, and setting off earthquakes.

FOLDING TECHNIQUES

ROCKS TEND to form in flat layers called strata. Some, called sedimentary rocks, form when sand and gravel and seashells settle on the sea bed. Volcanic rocks form as hot molten rock streams from volcanoes, that flood across the landscape.

Although rock layers may be flat to begin with, they do not always stay that way. Most of the great mountain ranges of the world began as flat layers of rock that crumpled. They were crumpled by the slow but immensely powerful movement of the great tectonic plates that make up the Earth's surface. Mountains occur mostly where the plates are crunching together, pushing up the rock layers along their edges into massive folds. These two experiments help you to understand just how that happens. Sometimes, the folds can be tiny wrinkles just a few centimetres long in the surface of a rock. Sometimes, they are gigantic, with hundreds of kilometres between the crests of each fold. As the layers of rock are squeezed horizontally, they become more and more folded. Some folds turn right over to form overlapping folds called nappes. As nappes fold upon each other, the crumpled layers of rock are raised progressively higher to form mountains. Sometimes you can see the complicated twists of folds on the side of a mountain or cliff.

Mountain range
Although many mountain ranges are getting higher at this very moment as plates move together, mountain building is thought to have been especially active at certain times in Earth's history known as orogenic phases, each lasting millions of years.

SIMPLE FOLDS
You will need: thin rug.

1 Find an uncarpeted floor and lay the rug with the short, straight edge up against a wall. Make sure the long edge of the rug is at a right angle to the wall.

2 Now push the outer edge of the rug towards the wall. See how the rug crumples. This is how rock layers buckle to form mountains as tectonic plates push against each other.

3 Push the rug up against the wall even more and you will see some of the folds turn right over on top of each other. These are folded-over strata or layers, called nappes.

PROJECT

COMPLEX FOLDS

You will need: *rolling pin, different colours of modelling clay, modelling tool, two blocks of wood measuring 5cm square, two bars of wood measuring 10cm x 5cm.*

1 Roll out the modelling clay into flat sheets, each about 0.5cm thick. Cut the clay into strips about same width as the blocks of wood. Square off the ends.

2 Lay the plasticine strips carefully one on top of the other, in alternating colours or series of colours. These strips represent the layers of rock.

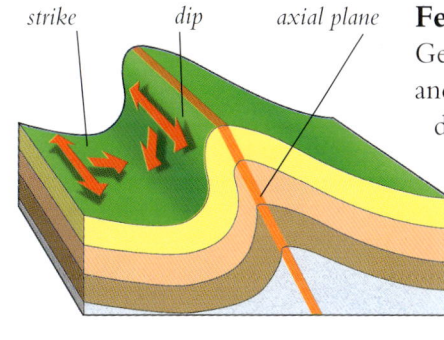

strike   dip   axial plane

Features of a fold

Geologists describe an upfold as an anticline and a downfold as a syncline. The dip is the direction the fold is sloping. The angle of dip is how steep the slope is. The strike is the line along the fold. The axial plane is an imaginary line through the centre of the fold – this may be vertical, horizontal or at any angle in between.

3 Place the blocks of wood at either end of the strips of plasticine. Lay the bars of wood down either side of the strips to stop them twisting sideways.

5 From time to time, stop and pull away the bars of wood to see what is happening. As you push harder, see how the layers crumple increasingly.

4 Ask a friend to hold on to one block while you push the other towards it. As you push, the effect is similar to two tectonic plates slowly pushing together.

WEARING AWAY

IF YOU went to the Moon, you would see the footprints of the first astronauts to set foot there way back in 1969. They are still there because the Moon's surface never changes. It has no atmosphere and its surface is now completely still. Earth's surface changes all the time. Most of these changes take place over millions of years, far too gradually for us to see. Occasionally, though the landscape is reshaped dramatically and quickly, such as when a volcano throws up a completely new mountain in minutes, or an avalanche brings down the entire side of a hill in seconds. The Earth's surface is shaped in two ways. First, it is distorted and reformed from below by the gigantic forces of the Earth's interior. Second, it is shaped from above by the weather, running water, waves, moving ice, wind and other agents of erosion.

The mountain's end

In mountain regions, the attack of the weather upon rock can break off huge quantities of debris. Millions of years' shattering, especially by frost, can create vast numbers of angular stones called "scree". As the stones fall off steep slopes, they gather at the foot of the slope in huge piles called scree slopes. Eventually, these stones will be broken down. They will form fine sand and silt and gradually wash away down to the sea where they can begin to form new rock.

Cycle of erosion

A century ago, Harvard Professor W. M. Davis (1850–1935) suggested that landscapes are shaped by "cycles of erosion" going through "youth, maturity and old age". This theory gave an idea of how landscapes evolve, but research has shown that the truth is more complex.

Youth: After an uplift of the land, there is vigorous erosion as fast-flowing streams bite deep into the landscape.

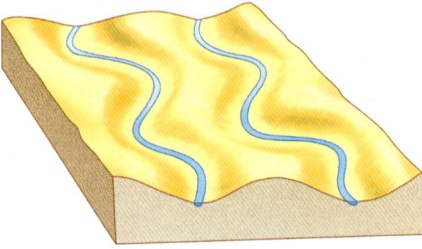

Maturity: River valleys get wider and slopes get gentler as they are worn away. Hill tops are rounded off.

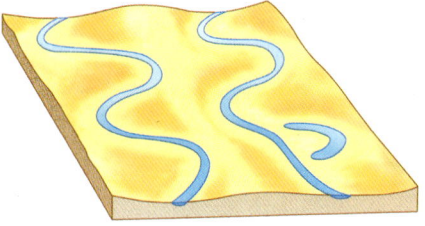

Old age: Valleys are worn flat into wide plains called peneplains and slopes are reduced to isolated hills.

Shattered peaks

As soon as rock is exposed to the weather, it starts to break down under the assault of wind and rain, frost and sun. Sometimes the rock is corroded (eaten away) by chemicals in the air, or rainwater trickling over it. Sometimes it is broken down physically by, for example, the effects of heat and cold. Water in cracks can expand so forcefully as it freezes that it can shatter the toughest rock.

Tors and kopjes

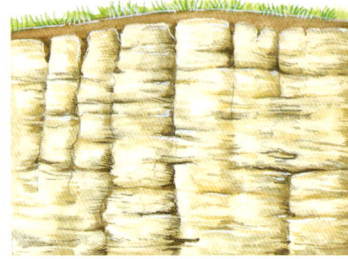

Water trickles down into the ground through joints and cracks in the rock.

The water corrodes and widens the joints, and the soft debris is washed away.

Eventually, only the big, intact blocks are left perched on the hill top.

In cool regions, there are outcrops called tors on the tops of moors. What is surprising about tors is that they are made from massive blocks that poke above completely smooth slopes, like a castle on a hilltop. There are similar outcrops called kopjes in the tropics. Both features are thought to have been created by the gradual uncovering of rock that has been corroded into big blocks by chemicals in water trickling through the ground, as the pictures above show.

Desert sculpture

In most places, running water is the main agent of erosion. The landscape is moulded into rounded hills and deep valleys by the wash of rivers and streams. In deserts, however, running water is sparse, although rivers may flow for a short time after rain and cut valleys. Much of the landscape is sharp and angular and cut into weird shapes by the blast of windblown sand. This is Mexican Hat in the Utah Desert in North America.

FACT BOX

• At -22°C, ice can exert a pressure of 3000kg on an area the size of a coin.

• The Colorado River began carving out the Grand Canyon 60 million years ago, as the river plain was slowly uplifted by giant movements of the Earth's crust, forming a plateau.

Acid work

A limestone statue on Gloucester Cathedral in England has worn away. Some rocks are better than others for building materials. All rain is slightly acidic. Limestone rock is especially susceptible to corrosion by acidic rainwater. The rainwater seeps into cracks in the rock, eating it away underground and creating potholes, tunnels and spectacular caverns.

PRACTICAL EROSION

Rocks and mountains look tough, but all the time they are being worn away by the weather, by running water, by waves, by glaciers and by the wind. It is such a slow process you can rarely see it happening. These three projects speed up the process so that you can see instantly what takes millions of years in Nature. The landscape is slowly reshaped as rocks crumble and mountains are worn down in a process called denudation (laying bare). Much of the damage is done by the weather.

Wherever rock is exposed to the weather, it is attacked by the atmosphere and begins to crumble away. This process is called weathering. Mechanical weathering is when rock is broken down by heat and cold. In areas where the temperature falls below freezing, water seeps into cracks in rocks and then freezes. It expands as it freezes, making the rock shatter or split. Chemical weathering is when the slight acidity of rainwater dissolves rock like tea dissolving sugar. Limestone landscapes can be corroded into fantastic shapes by chemical weathering.

Eroded landscape
Often, the effects of denudation are hidden under a covering of soil and debris. But they are obvious here in this dry landscape where the damage done by the weather and water running off the land is clearly exposed.

THE DESTRUCTIVE POWER OF WATER

You will need: baking tray, brick, tray or bowl, sand, castle mould, jug, water.

1 Put one end of a baking tray on a brick. Put the other end of the baking tray on a lower tray or bowl, so that it slopes downwards. Make a sandcastle on the baking tray.

2 Slowly drip water over the castle. Watch the sand crumble and form a new shape. This is because the sand erodes away where the water hits it.

3 Ensure the water flows down the centre of the baking tray. This way the water hits the middle of the sand castle, eroding the centre to form a natural stack.

HARD WATER

You will need: 3 small foil trays or saucers, jug, mineral water, tap water, distilled water, 3 labels, pen.

1 Fill three foil trays or saucers with a small amount of water – one with mineral water, one with tap water and one with distilled water.

2 Label the trays with the type of water in them and leave them somewhere warm and well-ventilated. Ensure they will not be disturbed for a few days.

3 Examine the trays once the water has evaporated. You will see that the distilled water has not deposited minerals because it does not contain any. Mineral water deposits only a few minerals. Tap water deposits vary depending on where you live. In hard water areas, tap water flows over rocks such as limestone and chalk, and deposits lime minerals. Water in soft water areas flows over rocks such as sandstone. This does not dissolve in water and leave a deposit.

distilled water

tap water

mineral water

CHEMICAL EROSION

You will need: baking tray, brown sugar, jug of water.

1 Build a pile of brown sugar on a tray. Imagine that it is a mountain made of a soluble rock (dissolves in water). Press the sugar down firmly and shape it to a point.

2 Drip water on your sugar mountain. It will erode as the water dissolves the sugar. The water running off should be brown, because it contains dissolved brown sugar.

RUNNING RIVERS

RIVERS PLAY a large part in shaping the landscape. Without them, the landscapes would be as rough and jagged as the surface of the moon. Tumbling streams and broad rivers gradually mould and soften contours, wearing away material and depositing it elsewhere. Over millions of years, a river can carve a gorge thousands of metres deep through solid rock, or spread out a vast plain of fine silt. Wherever there is water to sustain them, rivers flow across the landscape. They start high in the hills and wind their way down towards the sea or a lake. At its head, a river is little more than a trickle, a tiny stream tumbling down the mountain slopes. It is formed by rain running off the mountainside or by water bubbling up from a spring. As the river flows downhill, it is joined by more and more tributaries and gradually grows bigger. As it grows, its nature as well as its power change dramatically.

Thundering water

Waterfalls are found where a river plunges straight over a rock ledge and drops vertically. Typically they occur where the river flows across a band of hard rock. The river wears away the soft material beyond the hard rock and makes a sudden step down. This is Victoria Falls in Zimbabwe, formed where the Zambezi River suddenly drops about 100m into a deep, narrow chasm. The spray and the roar of the water have given the Falls the local name *Mosi oa Tunya,* the smoke that thunders.

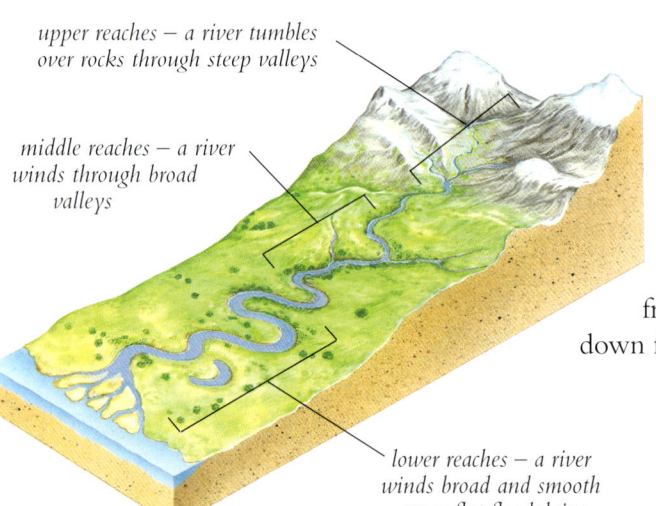

upper reaches – a river tumbles over rocks through steep valleys

middle reaches – a river winds through broad valleys

lower reaches – a river winds broad and smooth across flat floodplains

A river's course

As it flows down to the sea, a river changes its character. In the upper reaches, it is a fast-flowing, tumbling stream that cuts down through steep, narrow valleys. Lower down, a river broadens and deepens. Eventually, it meanders (winds) to and fro across broad floodplains made of material that it has washed down from higher up.

Forceful

High in the mountains, streams are small and tumble down over rocks and boulders. The valley is narrow with steep sides. Boulders often crash down into the stream bed so that the stream is forced to wind its way around them. The flow of water is very erratic. In flat places it flows slowly, whereas in other places it plunges down fast and furiously over rapids and waterfalls. Sometimes, when snow melts, the water level may rise enough to roll big boulders along.

Winding rivers

All rivers wind. As they near the sea, they wind more often, forming horseshoe-shaped bends called meanders, like these in the Guiana Highlands. Meanders begin as the river deposits sediments along its bed in ups and downs called pools (deeps) and riffles (shallows). The distance between pools and riffles, and the size of meanders, is usually in proportion to the width of the river. Meanders develop as the river cuts into the outer bank of a bend and deposits sand and mud on the inner bank.

Colorado loop

Big "gooseneck" bends usually only form when a river is crossing broad plains. But here on the Colorado, a gooseneck is in a deep gorge. A meander that cuts into a gorge in this way is called an incised meander. It probably formed millions of years ago when the Colorado Plateau was a flat lowland. The land was lifted upwards and the meander cut deeper as the land rose.

Black river

The Rio Negro, a major tributary of the Amazon in South America, is inky black. This is because of the rotting vegetable matter the river has picked up from the mangrove swamps it flows through. This is why it is called the Negro – *negro* is Spanish for black. Rivers carry their load of sediment in three ways. Big stones are rolled along the river bed. Smaller grains are bounced along the bed. The finest grains float in the water. Typically the load is mostly yellowish silt (fine mud) and sand. The Yangtse River in China carries so much yellow silt that it is often known as the Yellow River.

LAYING DOWN DEPOSITS

Sluggish waters
This river is flowing so slowly, as it nears the sea, that it cannot carry debris such as sand and mud. It begins to deposit its load, forming mudbanks, low-lying islands and fan-shaped deltas.

As rock is rotted away by the weather, the fragments move downhill. The rotted rock may slip down suddenly in a lump, creating a landslide or an avalanche. Sometimes, though, the debris creeps down, bit by bit, over many years. The rocky particles mix with vegetation and may be broken down into soil by further weathering and by bacteria and other organisms. Eventually the debris reaches the bottom of a slope. It is then washed away by a river, a glacier, by waves, or if the debris is very fine and dry, blown along by the wind. Big rivers such as the Mississippi carry huge quantities of debris. Even the biggest rivers, though, can only carry so much and must sometimes deposit their load of debris along the river's course. Rivers, glaciers, waves and wind all have their own characteristic ways of depositing debris as these projects show. Rivers flood out to deposit wide plains of silt called floodplains. Glaciers drop hummocks of material called moraines. Waves create beaches. Winds pile up dust in deposits called loesses or build up sand-dunes.

HOW WIND SORTS SAND

You will need: two empty ice cube trays, piece of card large enough to fit over an ice cube tray, spoon, mix of fine and coarse sand, hairdryer.

1 Turn one ice cube tray over, and lay it down end to end with another ice cube tray. Place the card over the upturned tray and spoon the sand over it to make a sand dune.

2 Hold a hairdryer close to the upturned tray, pointing towards the other tray. Turn the hairdryer on so that it blows sand into the open ice cube tray.

3 Look at the grains that have fallen into each box. The distance a grain travels depends on its weight. Heavy grains fall in the end of the tray nearest to you. Light grains are blown to the farthest end.

HOW WATER MAKES RIPPLES IN SAND

You will need: *heavy filled round tin, round plastic washing up bowl, water, fine clean sand, spoon.*

1 Place a heavy tin in the centre of the washing up bowl, then fill the bowl, with water to at least half way. The water should not cover more than two-thirds of the tin.

2 Sprinkle a little sand into the bowl to create a thin layer about ½cm deep. Spread the sand until it is even then let it settle into a flat layer at the bottom of the bowl.

3 Stir the water gently with the spoon. Drag it in a circle around the tin. As the water begins to swirl, stir faster, but keep it smooth.

Beach deposits

Waves often make ripples on a sandy beach like those in the project. Much of the rock debris washed down by rivers eventually ends up in the sea. Most settles slowly on the sea bed and compacts to form new sedimentary rocks. Some, though, is washed up by waves against the shore, and forms beaches.

4 As you stir faster, let the water swirl round by itself. The sand develops ripples. As you stir faster, the ripples become more defined.

ICE SCULPTING

GLACIERS ARE "rivers" of slowly moving ice that form in mountain regions where it is too cold for the snow to melt. They flow down mountain valleys, creeping lower and lower until they reach a point where the ice begins to melt.

The ice in a glacier is not clear, but opaque like packed snowballs. The surface is streaked with bands of debris, such as rocks, that has fallen from the mountain slopes above. Massive cracks appear where the ice travels over bumps in the valley floor. Today, glaciers form only in high mountains and near the North and South Poles. In the past, during cold periods called Ice Ages, they were far more widespread, covering huge areas of North America and Europe. When the ice melted, dramatic marks were left on the landscape. In the mountains of north-west America and Scotland the ice gouged out giant, trough-like valleys. Over much of the American Midwest and the plains of northern Europe lay vast deposits of till (rock debris).

Cracked ice
As a glacier moves downhill, it bends and stretches, opening deep cracks called crevasses. These may be covered by fresh falls of snow, making the glacier treacherous for climbers to cross. Crevasses may be a sign that the glacier is passing over a bar of rock on the valley floor. A bergschrund is a deep crevasse where the ice pulls away from the back wall of the cirque at the start of the glacier.

Old and new snow
The snow on these mountain peaks is likely to be quite different from that of the glacier below. Glaciers are made up of névé (new snow) and a compacted layer of old snow, or firn, beneath. All the air is squeezed out of the firn so that it looks like ice. The ice becomes more compacted over time, turning into thick white glacier ice, which begins to flow slowly downhill.

bergschrund

medial (middle) moraine

cirque

crevasses

snout

terminal (end) moraine

The course of a glacier
A glacier typically begins in a hollow high in the mountains called a cirque. It then spills out over the lip of the cirque and flows down the valley. If the underside of the glacier is "warm" (about 0°C), it glides in a big lump on a film of water melted by the pressure of ice. This is called basal slip. If the underside of the glacier is well below 0°C, it moves as if there were layers within the ice slipping over each other like a pack of cards. This is called internal deformation. Glaciers usually move in this way high up where temperatures are lower, and by basal slip farther down.

Deep digging

The fjords of Norway were made by glaciers that carved out deep valleys well below the current sea level. When the ice retreated, sea flooded the valleys to form inlets that in places are over 1000m deep. Glaciers may be slow, but their sheer weight and size gives them the immense power needed to mould the landscape. They carve out wide valleys, gouge great bowls out of mountains, and slice away entire hills and valleys as they move relentlessly on.

Moraine and drift

The grey bank across the picture above is a terminal (end) moraine. This is where debris has piled up in front of a glacier that has melted. The intense cold around a glacier causes rocks to shatter, and as the ice bulldozes through valleys, it shears huge quantities of rock from the valley walls. The glacier carries all this debris and drops it in piles called moraines. Melting glaciers deposit blankets of fine debris called glaciofluvial (ice river) drift.

Alaskan tundra

This chill landscape in Alaska is shaped by its periglacial climate (the climate near a glacial region). Winters are long and cold with temperatures always below freezing. In the short summers, ice melts only on the surface, and so the ground beneath is permafrost (permanently frozen). Water collects on the surface and makes the land boggy. As the ice melts, it stirs and buckles the ground beneath. As the ground thaws, then freezes again, cracks form, creating deep wedges of ice.

Glaciated valley

A wide, U-shaped valley in Scotland is left over from the Ice Ages, when glaciers were much more widespread. The last of the Ice Ages ended about 10,000 years ago. Valleys like this were carved by glaciers over tens of thousands of years. They are very different from the winding V-shape of a valley cut by a river.

DESERT LANDSCAPES

NOT ALL deserts are vast seas of sand. Some are rock-strewn plains. Others are huge blocks of mountains standing alone in wide basins, or just empty expanses of ice, as in Antarctica. All have the one common characteristic of being very dry. The lack of water makes desert landscapes very different from any others. Wind plays a much more important part in shaping them, because there is neither moisture to bind things together nor running water to mould them. In the desert, wind carves many weird and unique landforms, from sculpted rocks to moving sand-dunes. Very few places in the world are entirely without water, and intermittent (occasional) floods do have a dramatic effect on many desert landscapes. Then, instead of the rounded contours of wetter landscapes, steep cliffs, narrow gorges and pillar-like plateaus called mesas and buttes are formed.

Water in the desert

A valuable pocket of moisture has formed in the desert. There is often water beneath the surface of a desert, which may be left over from wetter days in the past. It may be water from wetter regions farther away, which has run down sloping rock layers beneath the desert. Occasionally, the fierce desert wind blows a hollow out of the sand so deep that it exposes this underground water.

parabolic dune

transverse dune

barchan dune

seif dune

wind direction

Sand-dune styles

In some deserts, such as the Sahara, there are vast seas of sand, called ergs, where the wind piles sand up into dunes. The type of dune depends on the amount of sand and changability of the wind direction. Crescent-shaped dunes called parabolic dunes are common on coasts. Ones with tails facing away from the wind are called barchans. These dunes creep slowly forward. Transverse dunes form at right angles to the main wind direction where there is lots of sand. Seifs form where there is little sand and wind comes from different directions.

Dune sea

In the western Sahara, giant sand-dunes hundreds of metres high have formed as a result of two million years of dry conditions. In places, the wind has piled up the dunes into long ridges called draa, that stretch far across the desert like giant waves in the sea. Each year, these ridges are moved 25m farther by the wind, as sand is blown up one side and rolls down the far side. Space probes have seen dunes like these on the planet Mars.

Monuments to water

Monument Valley in Utah, USA, is a spectacular example of what water erosion can do in dry places. The monuments are physical features called mesas. They are protected from water erosion by a cap of hard rock. The softer, unprotected rock between them has been washed away over millions of years. As water flows in channels rather than overland, there is nothing to round the contours, and cliffs remain sheer.

Occasional rivers

Rain is rare in the desert. What rain there is flows straight off the land and does not soak in, so streams rarely flow all the time. Instead, most streams flow only every now and then, and are said to be ephemeral or intermittent. In between wet periods, they leave behind dry beds called arroyos. In the Sahara Desert and the Middle East, rare rain torrents wash out narrow gorges called wadis. These are normally dry, but after rain may fill rapidly with water in a flash flood.

The power of the wind

Strong winds blow unobstructed across the desert, picking up grains of sand and hurling them at rocks. The sandblasting can sculpt rocks into fantastic shapes such as this rock arch. Satellite pictures have revealed parallel rows of huge, wind-sculpted ridges in the Atacama Desert in Chile and in the Sahara Desert in Africa. These yardangs are hundreds of metres high and dozens of kilometres long.

FACT BOX

• Summer temperatures in the Sudan Desert in Africa can soar to 56°C, hotter than anywhere else in the world.

• Because desert skies are clear, heat escapes at night and temperatures can be very low.

• Not all the world's deserts are hot. Among the world's biggest deserts are the Arctic and Antarctic, both of which have hardly more rain than the Sahara.

SEA BATTLES

COASTLINES ARE constantly changing shape. They change every second as a new wave rolls in and drops back again, and every six hours as the tide rises and falls. Over longer periods, too, coastlines are reshaped by the continuous assault of waves. They change more rapidly than any other type of landscape. Of all the agents that erode (wear away) the land, the sea is the most powerful of all.

Huge cliffs are carved out of mountains, broad platforms are sliced back through the toughest of rocks, and houses are left dangling over the edges of the land. Such examples are proof of the awesome force of waves. The sea is not always destructive, though. It builds as well. Where headland cliffs are being eroded by waves, the bays between may fill with sand – often with the same material from the headlands. On low coasts where the sea is shallow, waves build beaches and banks of shingle, sand and mud. How a coastline shapes up depends on the sort of material it is made of and on the direction and power of the waves.

White cliffs
Waves can quickly wear the land into sheer cliffs like those at Beachy Head in England. When the last Ice Age ended around 8,000 years ago, the sea level rose as the ice melted. Sea flooded over the land to form what is now the English Channel. The waves quickly began to slice away the land. The valleys were cut off so that they became just dips in the cliffline, while the hills became crests.

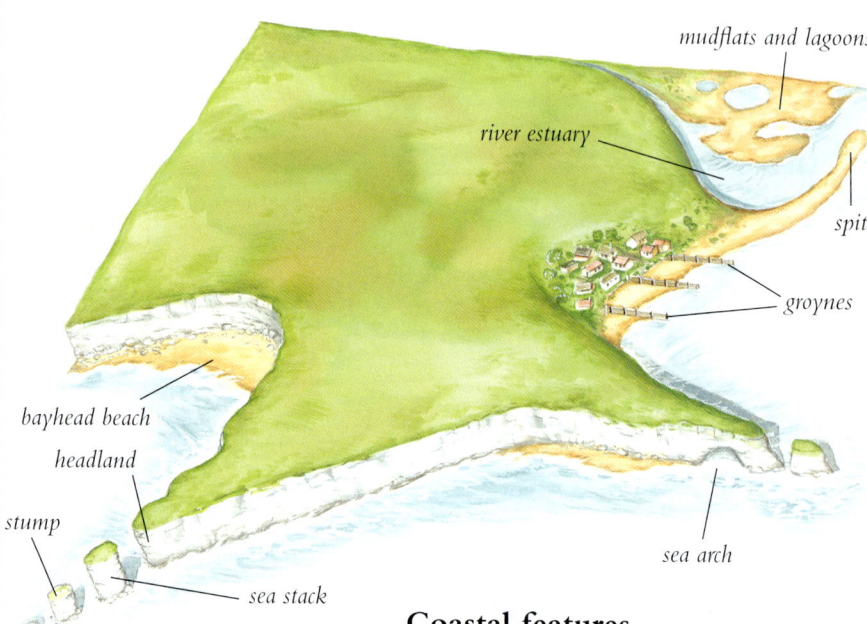

mudflats and lagoons

river estuary

spit

groynes

bayhead beach

headland

stump

sea stack

sea arch

Coastal features
The sea's power to build and destroy a coastline can be seen in this picture. Exposed parts of the coast that face into the waves are eroded into steep cliffs. Headlands are worn back, leaving behind stacks, stumps and arches. In more sheltered places, the sand piles up to form beaches, or waves may carry material along the coast to build spits and mudflats.

Saving the beach
Low fences called groynes have been built along this beach to stop it being carried away by the wave action. When waves hit a coast at an angle, they fall back down the beach at right angles. Any sand and shingle carried by the wave falls back slightly farther along the beach. In this way, sand and shingle is carried along the beach in a zigzag movement called longshore drift.

Storm force

During a storm, the waves crash on to the shore with tremendous force. The waves attack hard rock in two ways. They pound the rock with a huge weight of water filled with stones. They also split the rock apart as the waves force air into cracks. On high coasts, the constant attack of waves undercuts the foot of the slope, and unsupported upper parts topple down to create a cliff.

Wave–cut platforms

A pool has formed in a dip on a rocky platform on the seashore. The sea's erosive power is concentrated in a narrow band at the height of the waves. As the waves wear back sea cliffs, they leave the rock below wave height untouched. As the cliff retreats, the waves slice away a broad platform of rock. Geologists call this a wave-cut platform and it lies between the low-tide and high-tide marks. As the tide goes out twice each day, the sea leaves water in dips and hollows to form pools.

Rock arch

The sea arch at Durdle Door in Dorset, England, is made by waves eating away at large blocks of well-jointed rock. The waves have worked their way into joints in the rock and slowly enlarged them. Eventually, the cracks are so big that they open up into sea caves, or cut right through the foot of a headland to create a sea arch like this. When further erosion makes the top of an arch collapse, pillars called stacks are left behind. Pillars may then be eroded into shorter stumps.

PULLING TIDES

Every 12 hours or so, the sea rises a little in some places, then falls back again. These rises and falls are called tides, and they are caused mostly by the Moon. The Moon is some way away, but gravity pulls the Earth and Moon together quite strongly. The pull is enough to pull the water in the oceans into an egg shape around the Earth. This creates a bulge of water – a high tide – on each side of the world. As the Earth turns round, these bulges of water stay in the same place beneath the Moon. The effect is that they run around the world, making the tide rise and fall twice a day as each bulge passes. Actually, the continents get in the way of these tidal bulges, making the water slosh about in quite a complicated way.

The first experiment on these pages shows how the oceans can rise and fall a huge distance in tides without any change in the amount of water in the oceans at all. The second shows what the tidal bulge would look like if you could slice through the Earth, and how it moves around as the Earth turns beneath the Moon.

Low tide
The height of tides varies a great deal from place to place. In the open ocean, the water may not go up and down more than a metre or so. But in certain narrow inlets and enclosed seas, the water can bounce around until tides of 15m or more can build up.

HIGH AND LOW TIDE

You will need: old round plastic bowl (such as a washing up bowl), water, big old plastic ball to represent the world.

1 Place the bowl on a firm surface, then half fill it with water. Place the ball gently in the water so that it floats in the middle of the bowl.

2 Put both hands on the top of the ball, and push it down into the water gently but firmly. Look what happens to the level of water; it rises in a "high tide".

3 Let the ball gently rise again. Now you can see the water in the bowl dropping again. So the tide has risen and fallen, even though the amount of water is unchanged.

THE TIDAL BULGE

You will need: *strong glue, one 20cm length and two 40cm lengths of thin string, big plastic ball to represent the world, old round plastic bowl, water, adult with a simple hand drill.*

1 Glue the 20cm length of the string very firmly to the ball and leave to dry. Ask an adult to drill two holes in the rim of the bowl on opposite sides.

2 Thread a 40cm length of string through each hole and knot around the rim. Half fill the plastic bowl with water and float the ball in the water.

3 Ask a friend to pull the string on the ball towards him or her. There is now more water on one side of the ball. This is a tidal bulge.

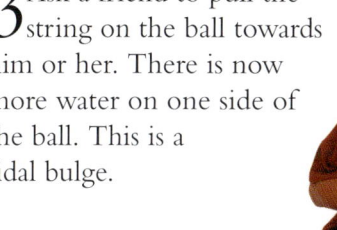

Pulling power

When the Moon and Sun line up at Full Moon and New Moon, their pulling power combines to create very high spring tides. A Half Moon means that the high tide will fall well below the highest tide mark. This is because the Sun and the Moon are at right angles to each other. Even though the Sun is farther away from Earth than the Moon, it is so big that its gravity still has a tidal effect. But at Half Moon, they work against each other and create the very shallow neap tides.

4 The Moon pulls on the water as well as the Earth. So now ask the friend to hold the ball in place while both of you pull out the strings attached to the bowl until it distorts.

5 There is now a tidal bulge on each side of the world. One of you slowly turn the ball. Now you can see how, in effect, the tidal bulges move round the world as the world turns.

THE OCEANS' HIDDEN DEPTHS

NEARLY THREE-QUARTERS of the world is under the oceans, lying an average of 3,730m under water. In places, the oceans plunge down to a depth of 11,000m, which is enough to drown the world's highest mountain, Mount Everest, and leave almost 2km to spare. Seas cover four-fifths of the Southern Hemisphere, and three-fifths of the Northern Hemisphere. The five great oceans around the world are the Pacific, the Atlantic, the Indian, the Southern (around Antarctica) and the Arctic. The biggest of them by far (although they are all actually linked together) is the Pacific, which covers almost a third of the Earth. Until quite recently we knew hardly more about the ocean depths than about the surface of Mars. However, in the last 40 years, there have been remarkable voyages in submersible craft capable of plunging to ever greater depths, and extensive oceanographic (ocean mapping) surveys. These have revealed an undersea landscape as varied as the continents, with mountains, plains and valleys.

Probing the depths

Knowledge of the ocean depths has increased dramatically, thanks to small, titanium-skinned submersibles and robot ROVs (remote-operated vehicles). These can withstand the enormous pressure of 3km of water above them. Satellites orbiting high above the Earth can make instant maps of the sea floor too. They pick up faint variations in the sea surface. These are created by changes in gravity, which in turn are caused by ups and downs in the ocean floor.

> ### FACT BOX
>
> • Seawater is 96.5% water and 3.5% salt. Most of the salt is sodium chloride (table salt).
>
> • There are canyons under the sea as big as the Grand Canyon.
>
> • The Mid-ocean Ridge is the world's longest mountain chain, winding 37,000km under three oceans, including the Atlantic.

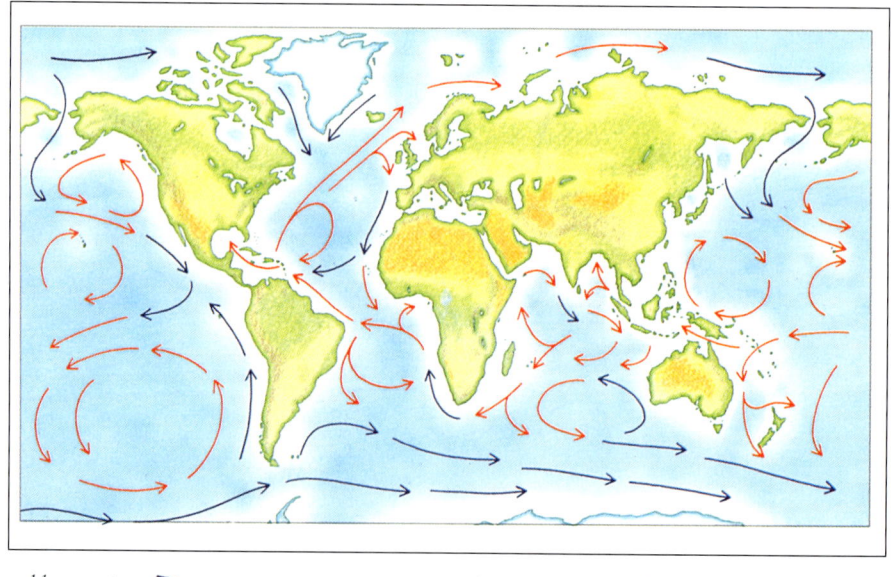

cold current ⟶   *warm current* ⟶

Currents on the ocean surface

The ocean waters are constantly circulating in currents. Those near the surface are driven along by the combined effect of winds and the Earth's rotation. They circulate in giant rings called gyres. In the Northern Hemisphere gyres flow clockwise, while in the Southern Hemisphere they flow anticlockwise. Deeper down, currents flow between the poles and the Equator. These are driven by differences in the water density, which varies according to the temperature and how salty the water is.

Island rings

The Maldives are a series of atolls – ring-shaped islands of coral – in the Indian Ocean. The coral ring first began to form around the peak of a seabed volcano that poked up above the sea's surface. At some time, the seabed moved. The volcano moved with it and slowly sank beneath the waves. The coral, however, kept on growing upwards, without its volcano centre. The reef is sometimes hundreds of metres deep.

Coral reefs

Over millions of years huge colonies of tiny sea animals called coral polyps build reefs (ridges) just below the surface of the sea. As each polyp dies its skeleton becomes hard. Colourful living polyps live on the skeletons of dead ones, so gradually layers of polyps build up and the coral reef grows bigger. Coral reefs support an extraordinary variety of marine life.

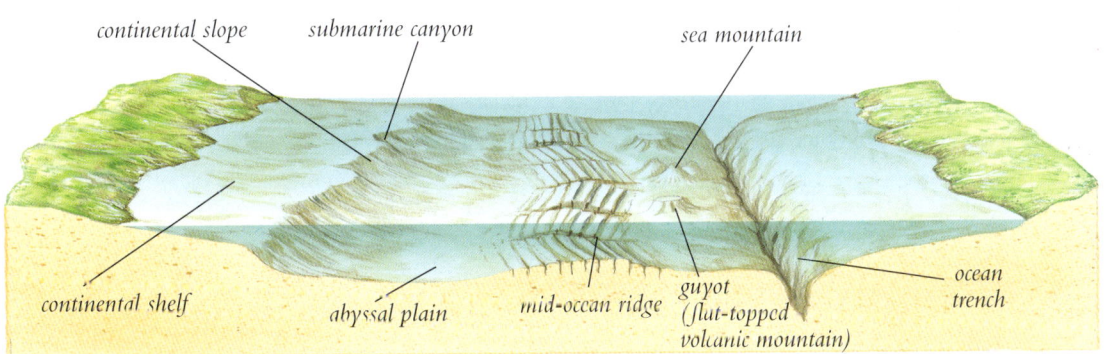

continental slope *submarine canyon* *sea mountain*

continental shelf *abyssal plain* *mid-ocean ridge* *guyot (flat-topped volcanic mountain)* *ocean trench*

The ocean floor

Running along the edges of each continent is a narrow shelf of shallow ocean barely 100m deep, called the continental shelf. Beyond, the ocean bed plunges down the continental slope into the depths of the ocean basin 2,000m below. The floor of the ocean basin is an almost flat plain, called the abyssal plain, which is covered with a thick slime called ooze.

Giant waves

The world's biggest waves occur in the biggest oceans, the Pacific and the Atlantic – and in the Southern Ocean around Antarctica. The further winds blow over the oceans, the bigger waves are likely to be. In the Southern Ocean, the winds roar right round the world unhindered by land. Monster waves estimated to be 40 or 50m high – as high as an apartment block – have been spotted from time to time.

OCEANS ON THE MOVE

You will need: *rectangular washing up bowl, jug, water, bath or inflatable padding pool.*

MAKING WAVES

1 Place the bowl on the floor or on a table. Choose a place where it does not matter if a little water spills. Fill the bowl with water almost to the brim.

2 Blow very gently over the surface of the water. You will see that the water begins to ripple where you blow on it. This is how waves are formed by air movement.

THE OCEANS are very rarely completely still. Even on the calmest day, little ripples play across the surface, or the water gently undulates. When the weather is stormy, giant waves higher than a house can rear up and crash down, turning the sea into a raging turmoil.

Waves begin as the surface of the water is whipped up into little ripples by wind blowing across the surface. If the wind is strong enough and blows far enough, the ripples build up into waves. The stronger the wind and the longer the fetch (the farther they blow across the water), the bigger the waves become. In big oceans, the fetch is so huge that smooth, giant, regular waves called swells sweep across the surface, and waves may travel thousands of kilometres before they meet land.

Waves usually only affect the surface of the water. The water does move at a deeper level, in giant streams called ocean currents, if the wind blows again and again from the same direction. Some deep ocean currents, moved by differences in the water's saltiness or temperature, can stir up the water right down to the ocean bed. The first project shows how waves are made. Currents such as the ones in the second project happen in the oceans on a much larger scale, and circulations or gyres such as this swirl round all the world's major oceans.

3 Fill the bath or pool with water. Blow gently along the length of the bath or pool. Blow at the same strength as in step 2, and from the same height above the water.

4 Keep blowing for a minute or so. Notice that the waves are bigger in the bath or pool, even though you are not blowing harder. This is because the fetch is bigger.

OCEAN CURRENTS

You will need: *rectangular washing up bowl, jug, water, talcum powder.*

1 Place the bowl on the floor or on a table. Choose a place where it does not matter if a little water spills. Fill the bowl with water almost to the rim.

2 Scatter a small amount of talcum powder over the water. Use just enough powder to make a very fine film over the surface. The less you use, the better.

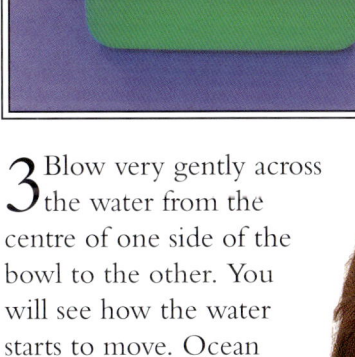

3 Blow very gently across the water from the centre of one side of the bowl to the other. You will see how the water starts to move. Ocean currents begin to move in the same way.

Wind and spin

Waves in the chill Antarctic oceans are driven by the wind, ocean surface and the effect of the Earth's rotation. The main winds in the tropics, called trade winds, blow the waters westwards along the Equator, creating equatorial currents. When these currents run into continents, they are deflected, still warm, towards the poles by the Earth's rotation. Eventually, they run into westerly winds which blow the water back eastwards again.

4 Keep blowing and the powder swirls in two circles as it hits the far side. This is what happens when currents hit continents. One current turns clockwise, the other turns anti-clockwise.

THE WORLD'S WEATHER

Storms, winds, snow, rain, sunshine, calm and all the other things we call weather are simply changes in the air. Sometimes these changes can happen very suddenly. A warm sunny day can turn to storm, bringing high winds and lashing rain that then end just as abruptly as they began. In some places, such as in the tropics, on either side of the Equator, there is very little difference in the weather from one day to another.

The planet's weather and every change in it is governed by the heat of the Sun. Winds blow up, for example, when the Sun heats some places more than others. This sets the air moving. Rain falls when air warmed by the Sun lifts moisture high enough for it to condense into big drops of water. On satellite photographs, swirls of cloud indicate how the air is moving. From them, meteorologists can identify distinct circulation patterns and weather systems, such as depressions and fronts, each of which brings a particular kind of weather.

The world's driest place

The Atacama Desert in Chile is the world's driest place, receiving little more than 1cm of rain in a year. It is dry because winds blow in from the Pacific Ocean over cold coastal currents. The cold water cools the air so much that all the moisture in it condenses before it reaches the land. So, as it blows over the Atacama, it is very dry.

Antarctic chill

The coldest place in the world is Antarctica. In Vostok, the Russian research station in Antarctica, the temperature averages –57.8°C, and once dropped to –88°C. The Sun strikes polar regions at a low angle, not from directly overhead as it does at the Equator, so its power is severely reduced. In winter, the sun is below the horizon for most of the time, and it is night in the polar region for three icy months.

Winds of the world

Some winds are local, others blow only for a short while. Prevailing winds are those that blow for much of the year. The map shows the world's three major belts of prevailing winds. Trade winds blow between the Tropics of Cancer and Capricorn on either side of the Equator. They are dry winds from the East. Moist westerlies from the West blow between the tropics and the polar regions. Icy polar easterlies blow around the North and South Poles.

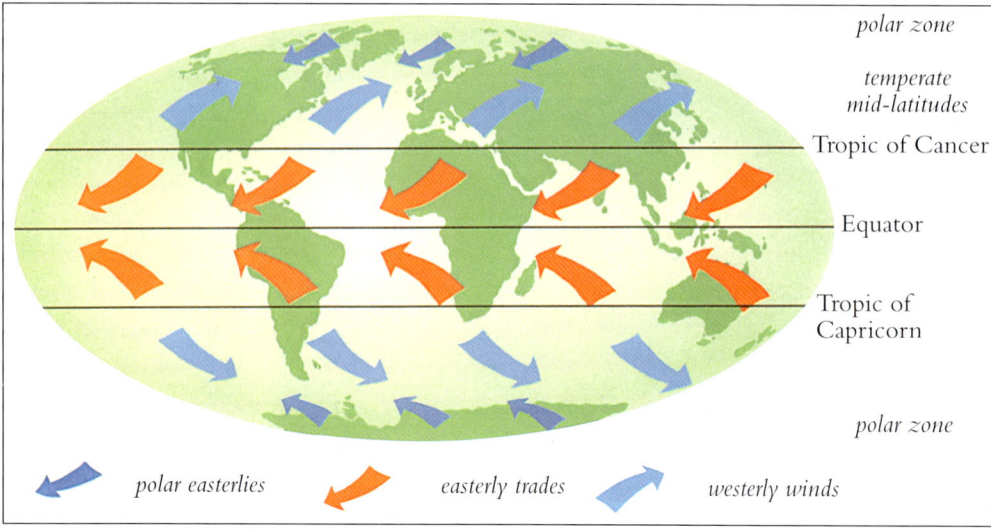

polar zone

temperate mid-latitudes

Tropic of Cancer

Equator

Tropic of Capricorn

polar zone

polar easterlies   easterly trades   westerly winds

Green and moist

The pasturelands of England are lush and green because they are well-watered by rain. England is in the westerly wind belt. Most of the winds blow in from the West over the Atlantic Ocean, where they pick up plenty of moisture that later falls as rain. Westerly winds also bring storms called depressions. These are places where winds spiral in towards a core of low pressure air. Depressions bring rain storms to the west coasts of Europe and North America as they move slowly east.

Perfect climate

Few places have such perfect weather as Quito in South America. It is close to the Equator, so the air is warm, but because it is high up, it never becomes too warm. The temperature in Quito never drops below 8°C at night, nor rises above 22°C during the day. The weather is made even more perfect by the fact that just 100mm of rain falls each month. It is no wonder, then, that the city of Quito is called the "Land of Eternal Spring".

FACT BOX

• The wettest place in the world is Tutunendo in Colombia, where the rainfall averages 1,170cm in a year.

• The hottest place is Dallol in Ethiopia, where it averages 34.4°C in the shade.

• The place with the most extreme weather is Yakutsk in Siberia. Here winters can be –64°C and summers 39°C.

Monsoon rains

For six months of the year, some parts of the tropics, such as India, are parched dry. Then, suddenly, torrential monsoon rains arrive as the summer sun heats the land. Air warmed by the land rises, and cool, moist air from the sea is drawn in underneath. This rain-bearing air pushes inland. Showers of heavy rain pour down during the wet season. Then, after about six months, the land cools and the winds reverse and blow out to sea. Immediately, the rain eases and the dry season is back.

CLIMATE CHANGE

EIGHTEEN THOUSAND years ago, the world was in the grip of bitter cold. A third of the planet was covered by thick sheets of ice. Vast glaciers spread over much of Europe and pushed far south into North America. This was just the most recent Ice Age. In the future, there will be another. The world's climate changes constantly, becoming warmer or colder from one year to the next, by the century or over thousands of years. These changes may be caused by a shift in the Earth's position relative to the Sun, or by bursts of sunspot activity in the Sun. Natural events such as volcanic eruptions, the impact of meteorites, or the movement of continents, also affect the weather. Recently, scientists have been concerned by the sudden warming of the world, triggered by air pollution.

Storms warn of global warming

Extra warmth from global warming puts extra energy into the air, bringing storms as well as warmer weather. The industrial world pumps huge amounts of gases into the air, including carbon dioxide from burning oil in cars and power stations. These greenhouse gases are so called because they trap the Sun's heat in the atmosphere like the glass in a greenhouse.

Sunspot storms

Sunspots are dark spots on the Sun where the surface is less hot. They seem to change all the time and reach a peak every 11 years or so. Measurements from the Nimbus-9 satellite show that the Sun gives the Earth less heat when there are fewer sunspots. Weather records show that when they reach their maximum level, the weather on Earth is warmer and stormier. The next sunspot maximum is in the year 2002.

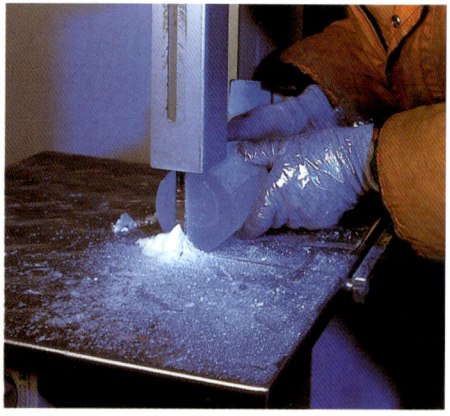

Ice core

A scientist investigates a core of solid ice that provides a remarkable record of climate change. The polar ice caps were built up over hundreds of thousands of years. Scientists drill into the ice of Greenland and Antarctica, and extract ice cores that are made up of layers of snow that have fallen over the years. They can detect changes in the atmosphere from microscopic bubbles of ancient air trapped within the ice and see how greenhouse gases have increased.

Polar ice caps

During Ice Ages, the Earth becomes so cold that the polar ice caps grow to cover nearby continents with vast sheets of ice. Ice Ages are periods of time lasting for millions of years. There have been four in the last billion years. During an Ice Age, the weather varies from cold to warm over thousands of years, and the ice comes and goes. There have been 17 glacials (cold periods) and interglacials (warmer periods) in the last 1.6 million years.

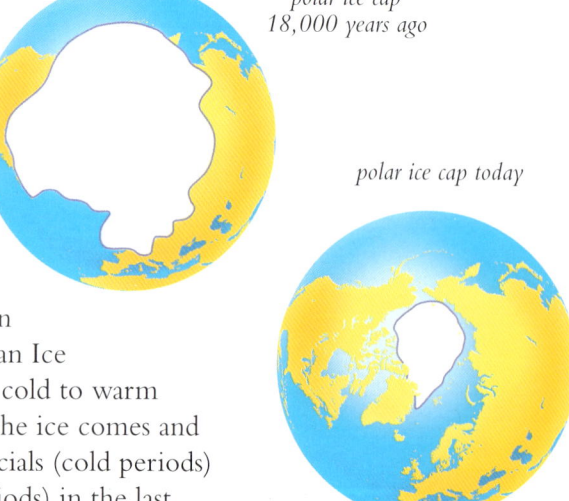

polar ice cap 18,000 years ago

polar ice cap today

Moving land

Everywhere on Earth has had a very different climate in the distant past. Fossils show that where New York now is was once a flaming desert, and icy Antarctica once enjoyed a tropical climate. This fossil is of a tropical fern, but it was found in Spitsbergen, which is well inside the Arctic Circle. Corals only survive in warm seas, but have been found in cold, northern seas. Such dramatic differences are not due to changes in the global climate, but because the continents have drifted around the globe. New York was once in the tropics. So was Antarctica.

Antarctic ice

The amount of ice in the world is fluctuating all the time. Antarctica contains 95 per cent of the world's ice and snow. But even Antarctica has not always been covered in ice. In fact, most Antarctic ice is less than ten million years old. The icicles form as the warmer weather comes and the ice begins to melt.

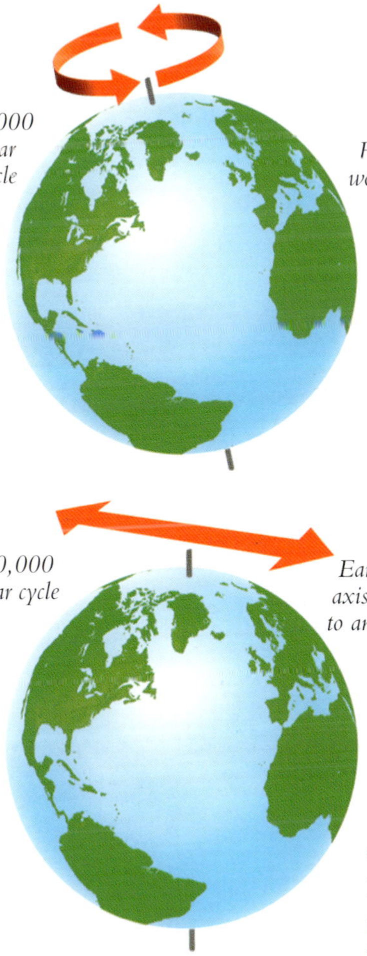

21,000 year cycle

Earth's axis wobbles round

Earth's orbit changing shape

Earth

Sun

40,000 year cycle

Earth's axis tilts to and fro

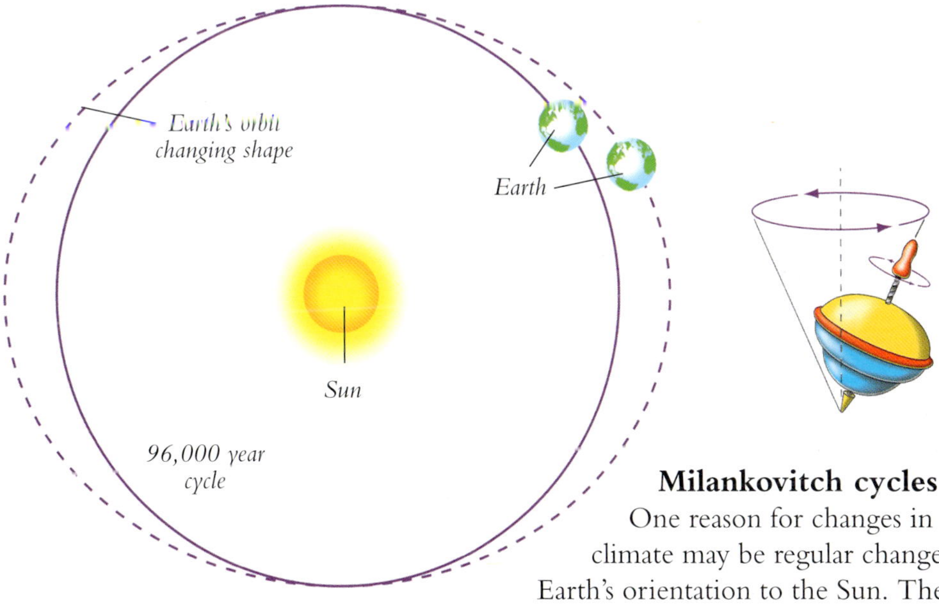

96,000 year cycle

Milankovitch cycles

One reason for changes in climate may be regular changes in Earth's orientation to the Sun. These are called Milankovitch cycles, after the Croatian scientist who discovered them. One cycle is the way Earth's axis wobbles like a top every 21,000 years. Another is the way its axis tilts like a rolling ship every 40,000 years. A third is the way its orbit gets stretched more or less oval over 96,000 years. All these changes affect how sunlight strikes the Earth – and so may have a dramatic effect on the Earth's climate.

WEATHER RECORD

Ancient wood

You may find a fallen or freshly sawn tree in a woodland to study. Scientists can take a record from a living tree by drilling out a small rod through the tree with a device called an increment borer. This gives a detailed record of climate changes Studying the past through tree rings is called dendrochronology, from the Greek words *dendron* (tree) and *chronos* (time).

CLIMATE IS the word used to describe the typical weather of a place at a particular time. The world's climate has swung this way and that throughout the planet's history. There have been times in the periods between Ice Ages, for example, when some plants and animals that are found in hot lands today lived in more northerly regions than they do now.

We can see how the climate has changed by studying weather records that have been made by people in the past, but these rarely go back more than 200 years. Scientists can also find many clues to climate change in Nature. In the sediments of sand and mud on ocean-beds, for instance, they found fossils of tiny shellfish called *Globorotalia*, which coils to the left in cold water and to the right in warm water. By working out when the sediments were laid down, the scientists could discover whether the water was warm or cold by the way the shellfish were lying.

This project shows how to make your own discoveries about recent climate change by looking at the year-by-year record of tree growth that is preserved in wood.

THE WOODEN WEATHER RECORD

You will need: *newly cut log, decorator's paintbrush, ruler with millimetre measurements, metric graph paper, pencil, calculator.*

1 Ask a tree surgeon, the local council or a sawmill for a newly cut slice of log. Use the paintbrush to brush away the dust and dirt from the slice of wood.

2 When the log slice is clean, examine it closely. Look at the pattern of rings. They are small in the centre and get bigger and bigger towards the outer edge of the log.

FACT BOX

• Each ring in the tree's cross-section represents a year's growth.

• The strong line at the edge of each ring marks the time in winter when growth stops.

• A wide ring indicates a warm summer with good growth.

• A narrow ring indicates a cool summer with poor growth.

• See if you can spot good summers and bad.

3 Each ring is a year's growth. So count the rings out from the centre carefully. This tells you how old the tree is. If there are 105 rings, for instance, the tree is 105 years old.

4 Using a ruler, measure the width of each ring. Start from the centre and work outwards. Ask a friend to write down the widths as you call them out.

Long-term calendar

All kinds of trees are useful for tree ring analysis, though the sequoias (redwoods) and pines of California are especially valued because some trees are over 4,000 years old. By comparing rings from different trees, scientists can build up a record of climate change.

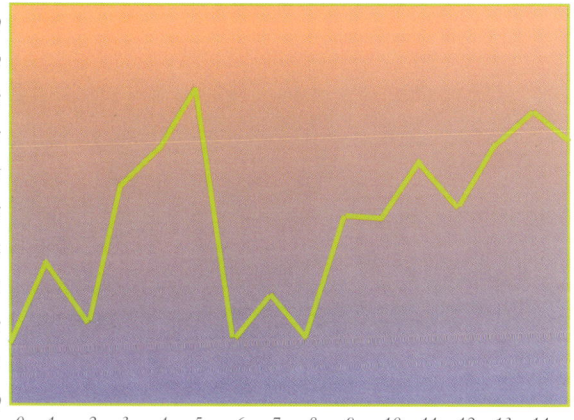

5 On graph paper, mark off the years along the bottom line, using five small squares for each year. Mark off widths for the rings up the side of the paper, using five small squares for each millimetre. Now, starting with year one on the left, plot your measurements as dots for each year across the graph.

6 Join the dots with a pencil line. This line shows how the weather has changed with each year. If the line is going up, the weather was warmer. If the line falls, the weather was colder. See if you can spot if it is getting warmer or colder over the years.

VEGETATION ZONES

SOME PLANTS, such as alpine grasses, can survive in very cold conditions, even if they are covered with thick snow for several months. Others, such as cacti, can cope with extreme heat. Each kind of plant thrives under particular conditions of soil and climate. Some groups of plants are so well adapted to conditions that exist in a particular region of the world that they are identified with those regions. The world can therefore be split into plant or vegetation regions according to the kind of plants that thrive there. Climate is the biggest influence on the kind of plants that grow, so vegetation regions tend to coincide with regions that have particular climates, such as tropical (near the Equator), or polar (near the poles). Many different plants live in each place, and within these broad regions, conditions can vary enormously.

Barren tundra

Only lichens, mosses, hardy grasses and tiny shrubs such as dwarf willows and birches grow in the tundra wastelands. In these polar regions, the temperature rarely rises above freezing, and then only in a few months of the year. Plants must survive on little or no water in winter because it is frozen. Then when the ice melts in spring, they have to cope with ground that is completely waterlogged.

Northern forests

Across the north of Russia and Canada are vast coniferous forests. This vegetation zone is called boreal forest or taiga. Winters are dark and cold, with thick snow. Conifers have thin, needle-like leaves that resist the cold, and snow falls easily off the cone-shaped trees.

Mixed woodland

In temperate regions, where summers are warm and quite moist, but winters are cool, the native vegetation is deciduous woodland. Deciduous trees lose their leaves in autumn. This reduces the need for water in winter, when frozen ground limits the water supply. Much of Europe and North America was once covered by vast deciduous woods, but over the centuries they have been cut or burned to make way for farmland.

Temperate grassland

Vast areas of the temperate zone, between the tropics and the poles, are covered with grass. Some of this has been created by farmers, as they have cleared woods for pasture, but much is natural. Temperate grassland is called by different names in different parts of the world, such as steppe in Asia and prairie in North America. Steppes are dry, so the grass is very short and coarse. Prairies are damper and the grass is lusher and longer.

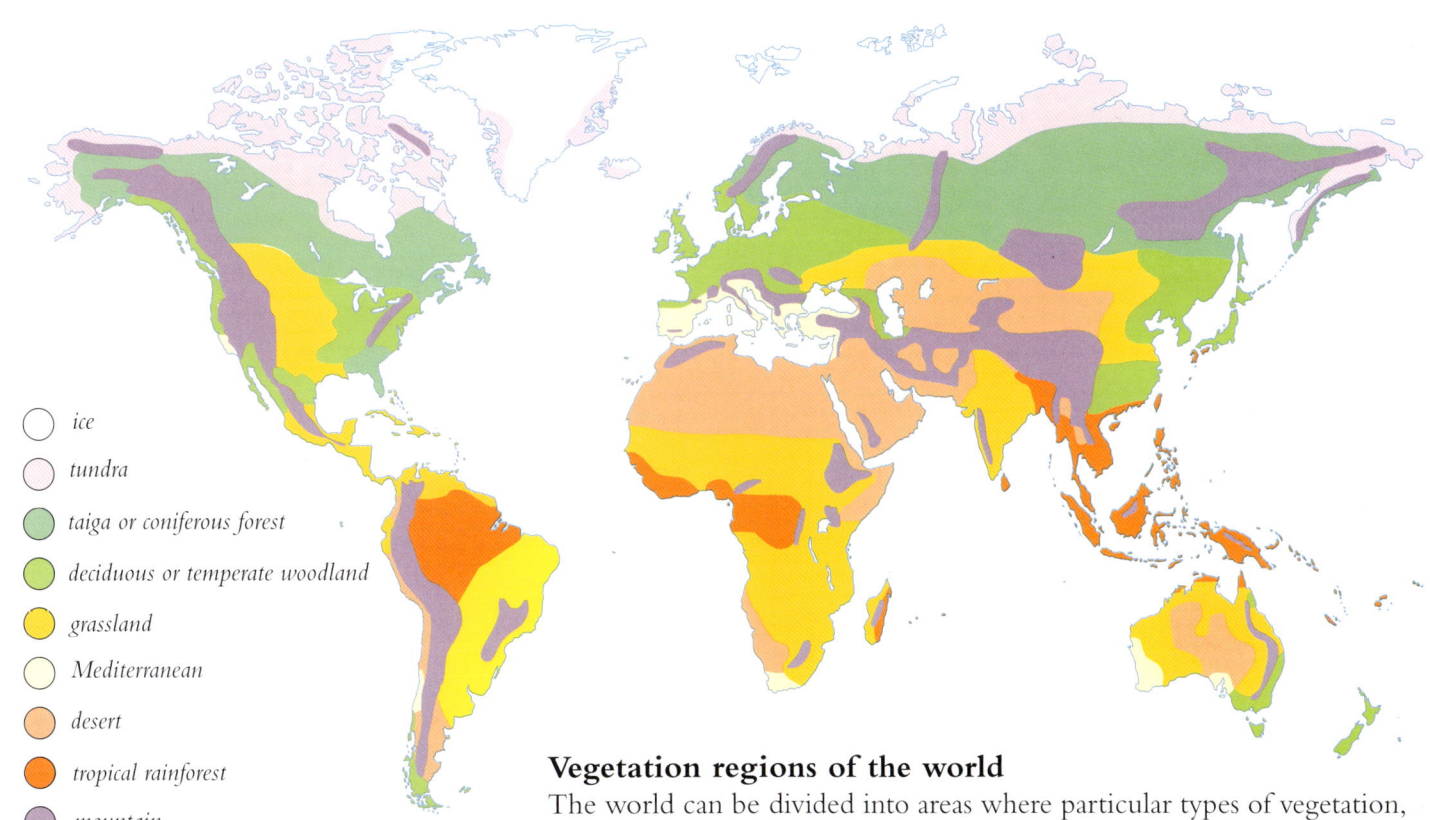

ice

tundra

taiga or coniferous forest

deciduous or temperate woodland

grassland

Mediterranean

desert

tropical rainforest

mountain

Vegetation regions of the world

The world can be divided into areas where particular types of vegetation, such as deciduous woodland, are dominant. The regions are closely linked to climate, as the type of vegetation that thrives depends on how warm it is or whether there is rain throughout the year or only in particular seasons. As the climate becomes colder away from the Equator towards the poles (and higher up mountains), there are fewer different types of plants.

Tropical rainforest

Warmth and plentiful rain year-round make tropical rainforests the richest plant habitats on Earth. Deciduous woods rarely have more than a dozen tree species, but tropical rainforests may have 100 or more in a single hectare. The forests are surprisingly fragile, because trees and soil are dependent on each other for survival. If trees are cut down, soil and the nourishing things it contains are quickly washed away.

Tropical grassland

Where rain in the tropics is seasonal, trees are rare as they cannot cope with the long, dry season. The typical vegetation is grassland, which in Africa is called savanna. Grasses in savanna lands grow tall and stiff. Dark evergreen trees such as acacias survive because their waxy leaves retain water and their thorns protect them from animals in search of moisture.

THE BALANCE OF LIFE

LIFE ON Earth may be classified into thousands of ecosystems. These are communities of living things that interact with each other and with their surroundings. An ecosystem can be anything from a piece of rotting wood to a huge swamp, but all the living things within it depend on each other.

Each living thing also has its own favourite place where factors such as temperature and moisture are just right. Some species can survive in a variety of habitats, but many can cope only with one. In an ecosystem, organisms depend on each other, and taking away just one species can threaten the existence of the others. If the plants on which a certain caterpillar feeds are destroyed, for example, the caterpillar dies, the birds feeding on the caterpillar starve and the foxes that feed on the birds go hungry, too.

Underwater richness
Coral reefs are the rainforests of the oceans. They provide shelter and food for an enormous range of marine plants and animals, from tiny coral polyps to giant clams and vicious predators. It is a fierce battle for life, food and space, however, and each species must develop its own programme for survival. Even the starfish and seasquirts in this picture are predators.

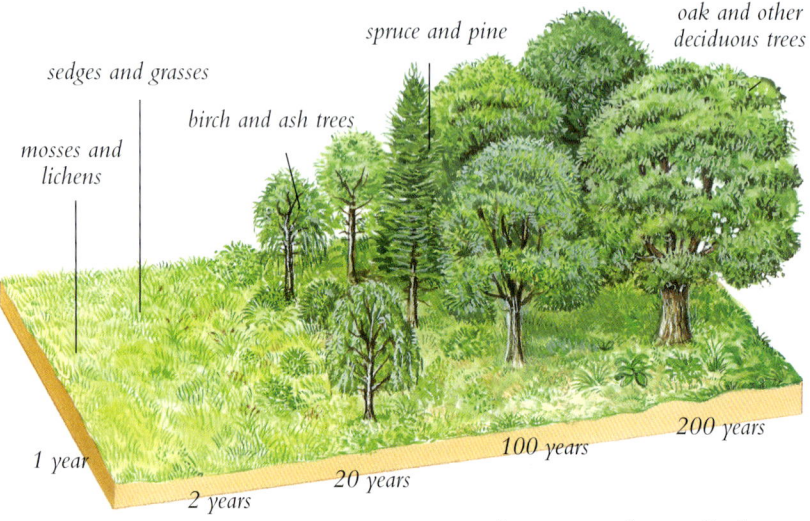

oak and other
deciduous trees

spruce and pine

sedges and grasses

birch and ash trees

mosses and
lichens

1 year

2 years

20 years

100 years

200 years

A succession of plants
When there is enough warmth and moisture on a piece of bare, rocky land, the first plants will grow. These will be the smallest, simplest plants, such as mosses and lichens, then tough grasses, that do not need very much to live on. The plants begin to hold the soil together. As they die and rot, they add nutrients to the soil, preparing it for bigger plants to grow. Soon there is enough to support small shrubs and tough trees such as pines, and eventually deciduous trees such as the oaks. This process is called vegetation succession, and it would take about 200 years for deciduous woodland to develop from the moss and lichen stage.

Harvest time
Farmland such as this has destroyed the natural vegetation and ecosystems. The number of plant species has dramatically reduced. A forest with hundreds of different plants may have been cleared to make way for a single crop. Because farming interrupts the flow of nutrients between soil and plants, the soils quickly become depleted, so farmers add artificial fertilizers.

Feeding habits

All animals depend on other living things for food and form part of an endless chain. This picture shows how the chain or food web works. A grasshopper eats a leaf, a thrush may eat the grasshopper and a kestrel may eat the thrush. When the kestrel dies and falls to the ground, bacteria break its body down and add nutrients to the soil so that new plants can grow. Herbivores eat plants only. Carnivores are meat eaters, and omnivores eat both vegetable and animal matter. Plants and algae make their own food from sunlight, and so are called autotrophs (self-feeders).

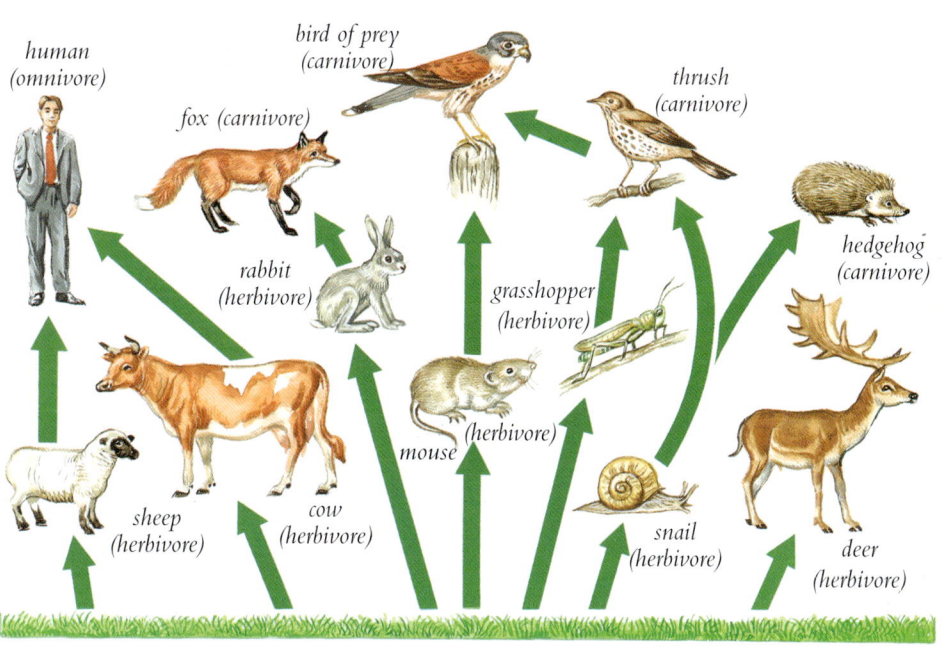

human (omnivore)

bird of prey (carnivore)

fox (carnivore)

thrush (carnivore)

rabbit (herbivore)

hedgehog (carnivore)

grasshopper (herbivore)

mouse (herbivore)

sheep (herbivore)

cow (herbivore)

snail (herbivore)

deer (herbivore)

emergents (trees rising above the main forest)

canopy (roof)

understorey

Layered jungle life

Different types of plant and animals inhabit different levels of the tropical forest. Isolated trees shoot up to emerge above the dense, leafy roof of the main forest. Some are 60m tall. Below is a dense canopy of leaves and branches on top of tall, straight trees, 30–50m tall. In the gloomy understorey beneath, young trees and shrubs grow and clinging lianas (climbing plants) wind their way up the trees.

A wealth of natural life

Swamps, ponds, and all the other places known together as wetlands, were once seen as useless land that could not be farmed or built on. More than half the wetlands in the United States have been drained in the last 100 years or so. However, wetlands are remarkably rich environments, producing up to eight times as much plant matter as the average wheat field. They can also play an important role in controlling floods and provide a valuable water store in times of drought.

FACT BOX

• Tropical rainforests cover less than 8 per cent of the Earth's land surface.

• They make up half of the world's growing wood and provide a home for 40 per cent of plant and animal species.

LIVING TOGETHER

THE WHOLE living world is a vast and everchanging jigsaw of plant and animal life. Each organism that is part of this living jigsaw links or interacts with other living things, either directly or indirectly. The whole picture is so huge and complicated that even for scientists, it is difficult to understand how it all works together all at once.

To make sense of it all, ecologists often break the living world down into lots of smaller units, such as tropical rainforests or freshwater lakes. Then they might break it down further into smaller regions, such as a mountain slope. They might go further still to identify individual trees, or a pool on a rocky seashore. Each of these units, where the things living there interact with each other, is called an ecosystem. One way in which the plant and animal life interacts is through food webs and chains, which show what eats what in an ecosystem. Warmth and shelter, protection from predators are other ways in which plants and animals can benefit from each other by co-existing in an ecosystem.

Arrested life

The axolotl from the lakes near Mexico City never grows up into a tiger salamander. It stays a tadpole all its life. This is because the water it lives in lacks iodine, the vital ingredient to make it grow. If the axolotl is given iodine injections, it turns into a tiger salamander. But that would alter the whole balance of life in the lakes. This shows how delicate the relationship is between each living thing and its environment.

MAKING YOUR OWN AQUARIUM ECOSYSTEM

You will need: *gravel, net, plastic bowl, water, jug, glass aquarium tank, rocks and lumps of old wood, water plants, jugful of pondwater, water animals.*

1 Put the gravel in a net and rinse in a plastic bowl of water or run it under the coldwater tap in the sink. This will discourage the formation of green algae.

2 Spread the gravel unevenly over the base of the tank to a depth of about 3cm. Add rocks and pieces of wood. These give surfaces for the snails to feed on.

PROJECT

3 Fill the tank to about the halfway mark with tap water. Pour the water gently from a jug to avoid disturbing the landscape and churning up the gravel.

4 Add some water plants from an aquarium centre. Keep some of them in their pots, but take the others out gently. Then root them in the gravel.

5 Now add a jugful of pondwater. This will contain organisms such as *daphnia* (water fleas), which add to the life of your aquarium. You can buy pondwater in a garden centre.

6 Now add a few water animals you have collected from local ponds, such as tadpoles in frog spawn or water snails. Take care not to overcrowd the aquarium.

Specialized living

This giant turtle lives on the Galapagos Islands. It was here that the ecologist Charles Darwin noticed how island plants and animals adapt to their local environment in quite distinctive ways because they are isolated. Although the turtles originally had no natural enemies, they are now threatened with extinction.

7 Place the tank in a reasonably bright light, but not in direct sunlight. You can watch the plants in the tank grow, Keep the water clean by removing dead matter off the gravel every 6 weeks.

HUMAN IMPACT

HUMANS NOW dominate the Earth to a greater extent than any other species of animal has ever done before. The Earth seems to be in grave danger of suffering irreparable harm from our activities. The demands that humans make on the planet so that they can feed themselves and live in comfort damage the atmosphere, the earth itself, and plant and animal life. Car exhausts and factory chimneys choke the air with pollution. Gases from supersonic jets and refrigerator factories make holes in the atmosphere's protective ozone layer. Rivers are poisoned by agricultural and industrial chemicals. Unique species of plants and animals vanish forever as their habitats are destroyed. Forests are felled, vast areas of countryside are buried under concrete, and beautiful marine environments are destroyed by tourism and sea traffic. The problem is not new, but as the pace of economic development increases, it is becoming more and more urgent to halt the destruction.

From forest to wasteland

A hillside that was once rich tropical forest has been slashed, burned and bulldozed. Vast areas of rainforest are being destroyed, in Brazil and Indonesia especially, to provide wood and to clear the land for rearing cattle. Unprotected soil soon turns to dust in the tropical sun, and farmers move on to wreak destruction on fresh forest.

Poisoned air

Cars, factories and homes pour fumes into the atmosphere and are making the air increasingly poisonous to humans and plants. Lead has been cut in car fuels because of the damage it was causing to children's brains, and other substances may be responsible for a rise in lung diseases. Burning fossil fuels adds sulphur dioxide to the water vapour in the air and causes acid rain, which pollutes lakes and kills trees, like these on Smokey Mountain, USA.

Deadly algae bloom

A choked and lifeless river such as this is common. Few rivers in the world are entirely free from pollution. Of 78 rivers tested in China, 54 were badly polluted with sewage and factory waste. In Europe, most rivers have high levels of nitrates and phosphates from chemical fertilizers washed off farmland. Heavily manured land can make nearby streams so rich in organic matter that algae multiplies and chokes all other life.

Life-saver

The rosy periwinkle is a tiny plant native to Madagascar. It was found to contain a chemical that has raised chances of children surviving leukaemia from 10 to 95 per cent, by preventing cell division. Each hour, about 2,400 hectares of the world's rainforests are destroyed. Much will include precious plants like this whose value we will never know.

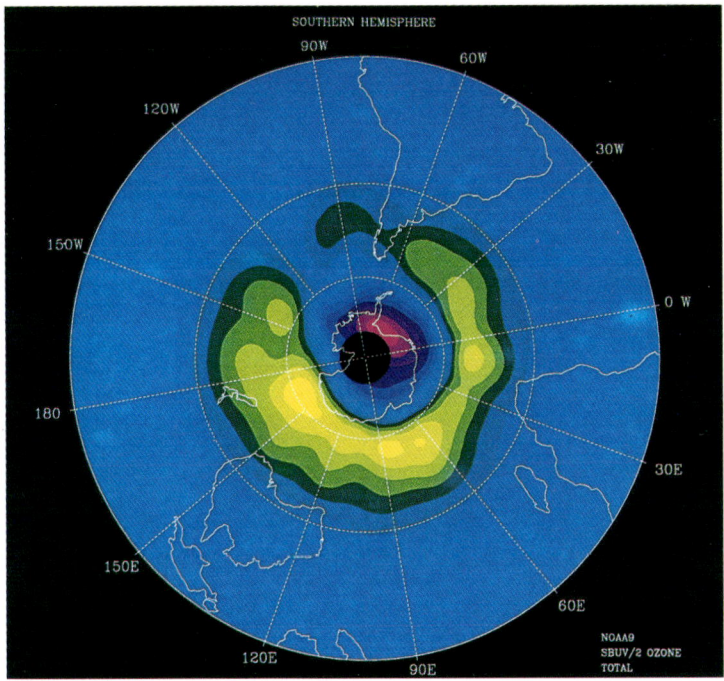

SOUTHERN HEMISPHERE

Destroying natural sunblock

The yellow and green blob on this satellite image of Antarctica is a huge hole in the ozone layer. Life on Earth depends on tiny amounts of the gas ozone in the stratosphere, 12 to 50km above the Earth. The ozone layer is our natural sunblock, and without it we have nothing to shield us from deadly ultraviolet (UV) rays from the Sun. This ozone hole reappears every spring, at both poles, each time getting bigger and staying for longer.

Exhaust fumes

Motor-vehicle exhausts discharge a huge range of unpleasant chemicals. Unburned hydrocarbons (better known as soot) make everything dirty and can cause breathing problems, while carbon dioxide adds to the greenhouse effect. Secondary pollutants are formed when exhaust mixes in the air. The worst of these may be ozone, which forms when sunlight makes soot react with other exhaust gases. Ozone is a good sunblock in the atmosphere, but dangerous when inhaled.

Endangered

The snow leopard is just one of millions of animal and plant species threatened with extinction by hunting or loss of natural habitat. Many species became extinct naturally, as the climate changed, or a food source ran out. Today the rate of extinction is 400 times faster than the all-time average, all due to human interference.

The Greenhouse Effect

Carbon dioxide in the air is important because it helps to trap warmth from the Sun, like the panes of glass in a greenhouse. In the past, this Greenhouse Effect has kept the Earth nicely warm. However, burning fossil fuels such as coal and oil have increased levels of carbon dioxide dramatically, and this is collecting around the Earth. Vital waves of heat radiated from the Sun can filter through this layer to warm the Earth. But heat waves generated on Earth are becoming trapped. They hit the carbon dioxide barrier and bounce back again. This is making the Earth warmer. Experts think temperatures will go up 4°C in the next 100 years, bringing extremes of weather, rising sea levels and flooding.

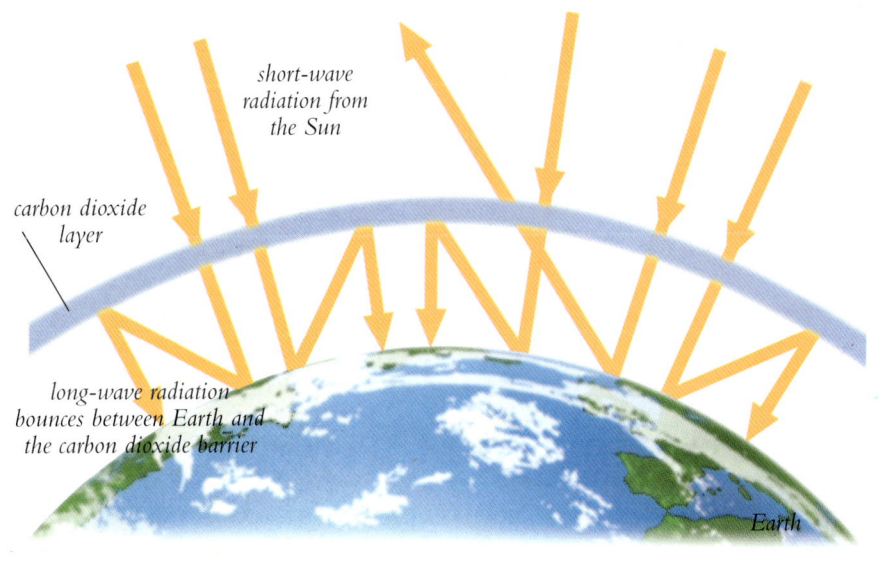

short-wave radiation from the Sun

carbon dioxide layer

long-wave radiation bounces between Earth and the carbon dioxide barrier

Earth

DANCE OF THE CONTINENTS

ONE OF the most amazing scientific discoveries of the 1900s was the idea of continental drift. Scientists discovered that the world's continents are not fixed in one place but are drifting slowly around the world – sometimes meeting, sometimes breaking apart. Recent high-precision measurements by satellite show that the continents are moving even now at between 2 and 20cm a year – about the pace of a fingernail growing. This may seem slow, but over the hundreds of millions of years of Earth's history, the continents have moved huge distances. There are ancient magnetic rocks within them that are like frozen compasses. Scientists can use them to plot how the alignment of the rocks have changed, and how the continents have twisted and turned. Piecing together these and other clues has gradually revealed just how the continents have moved over the last 750 million years.

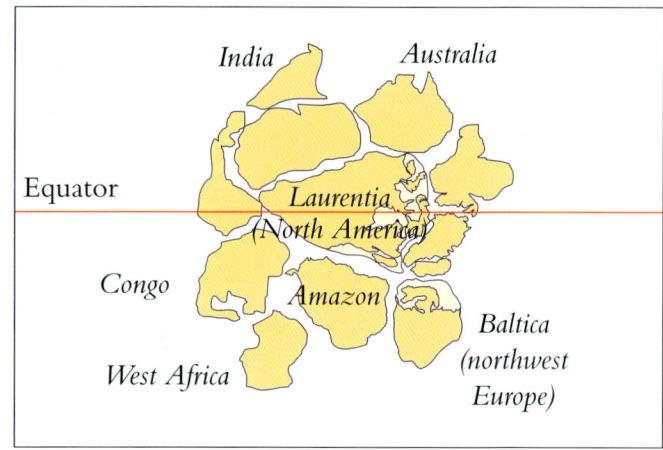

700 million years ago (mya)

All the continents are welded together in one giant continent, that today's geologists call Rodinia. There are none of the recognizable shapes of today's continents. Magnetic clues in the rocks confirm that North America lay at the continent's heart along the equator and northwest Europe to the south.

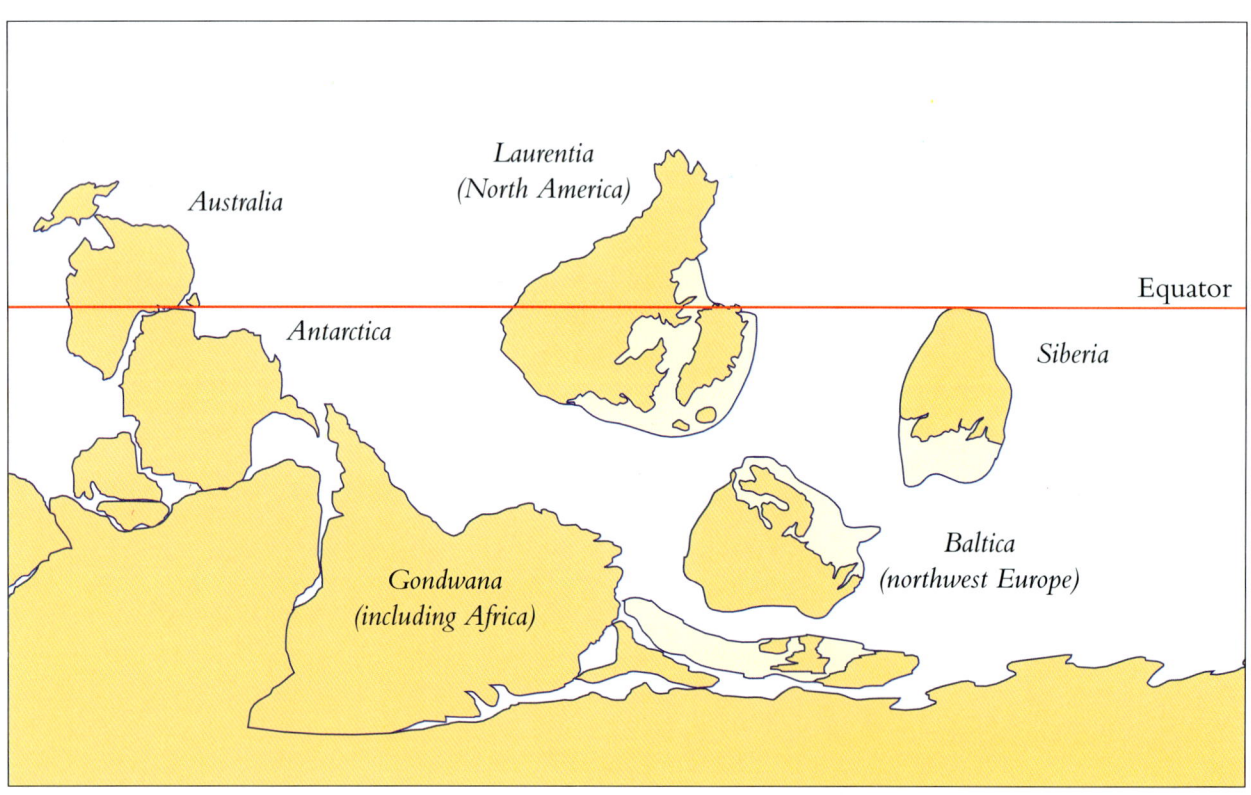

500 million years ago

The simple continent of Rodinia has broken up, but some of the fragments have gathered again around the South Pole. The map projection exaggerates the size of this South Pole continent, called Gondwanaland, but it was still massive, including all of today's Antarctica, Australia, South America, Africa and India.

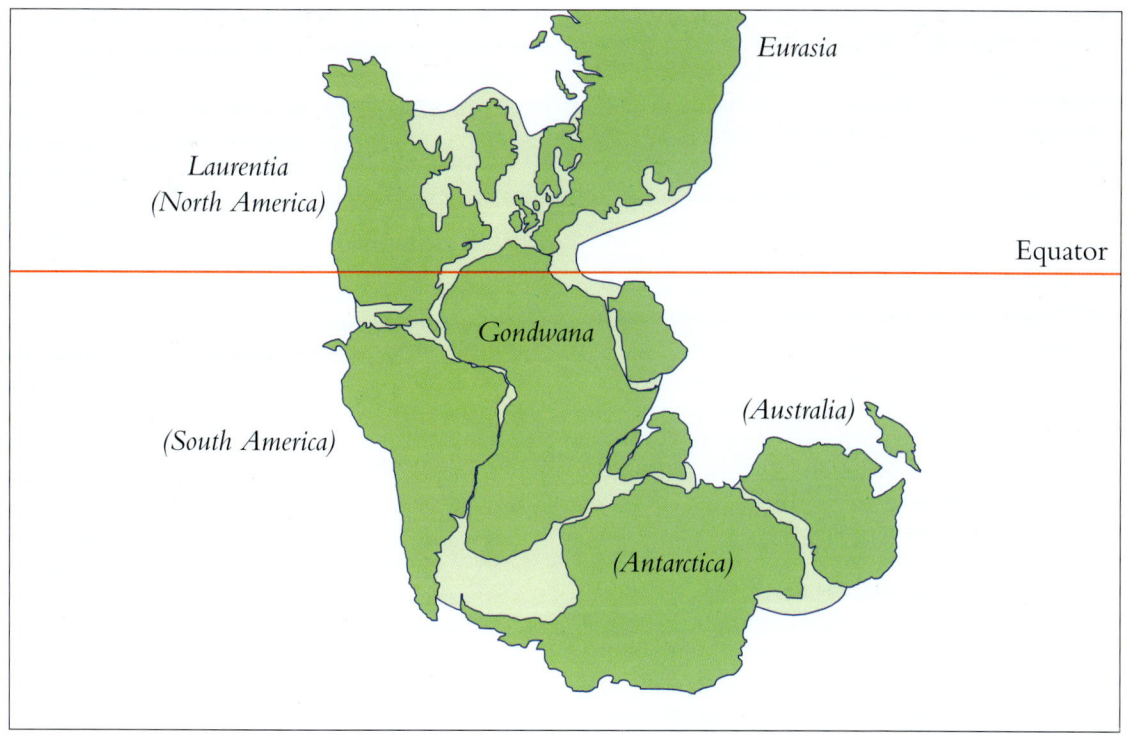

250 million years ago

Gondwanaland has merged with Laurentia and all the other continental fragments. A single mega-continent has formed and sits astride the equator. Geologists call this super-continent, Pangea (all Earth). Pangea was surrounded by a single ocean, which geologists call Panthalassa (all sea). 200 mya, soon after the dinosaurs first appeared on Earth, Pangea began to break up.

50 million years ago

Between 200 and 50mya, Pangea slowly broke up. First the Tethys Sea between Eurasia and Africa was opened up. Then the land split apart between Africa and South America to open into the South Atlantic ocean. By 50mya, North America has drifted away from Europe to open up the North Atlantic. India is powering north into southern Asia. Australia is out on its own.

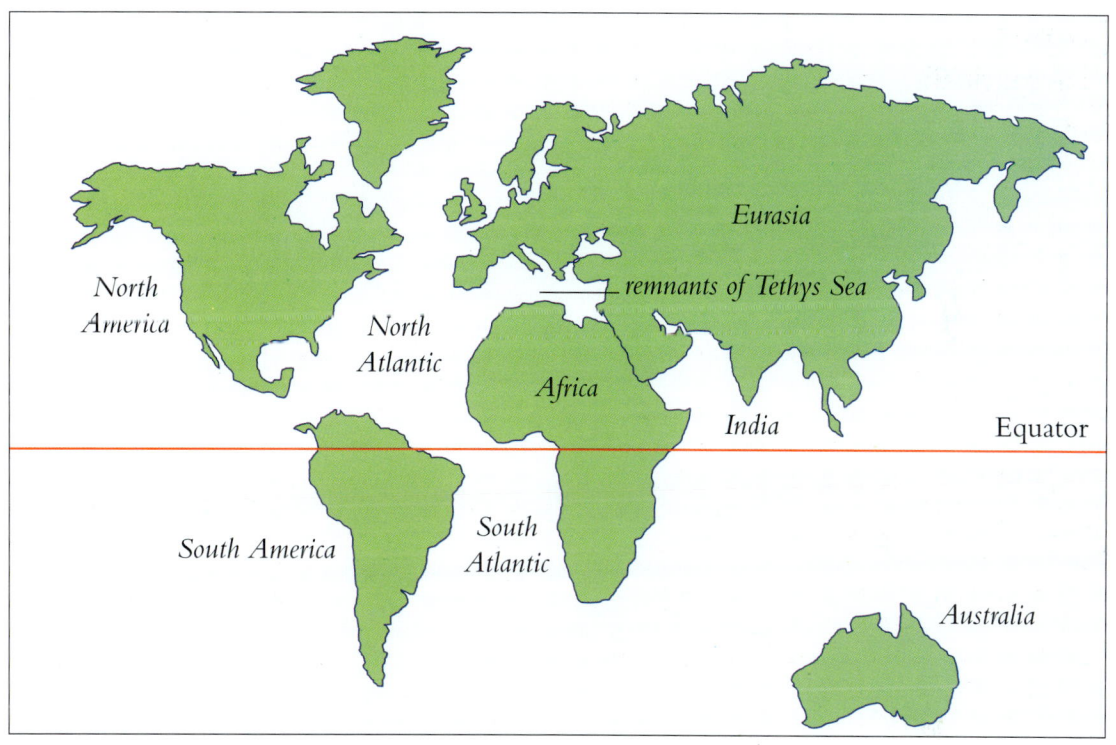

The world today

The continents today look familiar and fixed. But they are moving even now. In another 100 million years time the map of the world will look very, very different. The Americas are moving so far west that they will probably bump into eastern Asia in time, obliterating the Pacific. Africa will split into two parts and its East will drift into southern Asia. As for the rest of Earth, only time will tell.

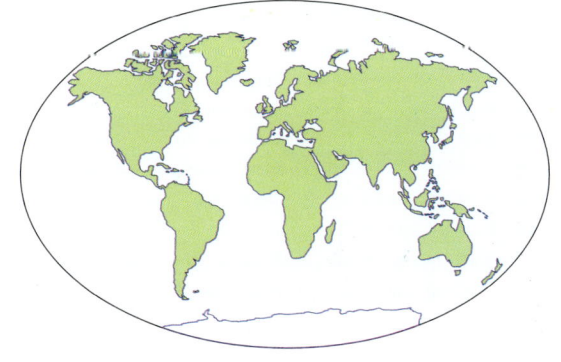

ROCKS AND MINERALS

These are the raw materials of the Earth's surface,
such as precious metals, sand, clay and soil.
Every valley, hill and mountain
is made of rocks.
Minerals form naturally in the earth.
Some rocks are just a few million years old.
Others are almost as old as the Earth itself.
Each rock is a clue
to the Earth's history, and studying
rocks helps us understand
the remarkable story of our planet.

Authors: Jack Challoner
and Rodney Walshaw
Consultants: Sue Bowler
and Bob Symes

ROCKS AND MINERALS

Rocks and minerals are the naturally occurring materials that make up planet Earth. We can see them all around us – in mountains, cliffs, river valleys, beaches and quarries. Rocks are used for buildings and many minerals are prized as jewels. Most people think of rocks as hard and heavy, but soft materials, such as sand, chalk and clay, are also considered to be rocks.

Minerals make up a part of rocks in the way that separate ingredients make a fruit cake. About 3,500 different minerals are known to exist but only a few hundred of these are common. Most minerals are solid but a few are liquid or gas. Water, for example, is a liquid mineral and many other minerals are found in liquid petroleum.

You might think that rocks last for ever, but they do not. Slowly, over thousands, even millions, of years they are naturally recycled, and the minerals that occur in a rock are moved from one place to another and form new rock.

Gemstones

Minerals prized for their beauty and rarity are called gemstones. The brilliant sparkle of a diamond appears when it is carefully cut and polished (as above).

pyrite

kyanite

copper

Minerals

All rocks are made up of one or more minerals. Minerals are natural, solid, non-living substances. Five different minerals are shown here. Each one has definite characteristics, such as its shape and colour, that distinguish it from all other minerals. Many types of minerals are found in thousands of different types of rock.

yellow sulphur crystals growing on kaolin

opal in ironstone

The Earth's crust

The Earth's surface is a thin, hard rocky shell called the crust. There are two kinds of crust – oceanic crust (under thc oceans) and continental crust (the land). The recycling of the rocks that form the crust has been going on for 4,000 million years.

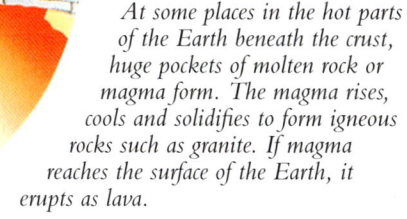

granite

sandstone

At some places in the hot parts of the Earth beneath the crust, huge pockets of molten rock or magma form. The magma rises, cools and solidifies to form igneous rocks such as granite. If magma reaches the surface of the Earth, it erupts as lava.

Sedimentary rocks, such as this sandstone, form from the fragments of other rocks that have been broken down by the action of rain, snow, ice and air. The fragments are carried away by wind or water and settle in a different place.

gneiss

Sometimes, within the Earth, the heat and pressure become so strong that the rocks twist and buckle and new minerals grow in them. The new rocks are called metamorphic rocks. This gneiss is a good example.

Fossils

As the fragments of rock settle in their eventual resting place, they may bury the animals and plants that lived there. The remains then become preserved as fossils. This ammonite was once a living creature, but is now made entirely of minerals.

Crystals

Minerals usually grow in regular shapes called crystals. When mineral-rich water fills a crack or cavity in a rock, veins and geodes may form. A geode is a rounded rock with a hollow centre lined with crystals. The beautiful crystal lining is revealed when it is split open. Geodes are highly prized by mineral collectors.

Getting at rocks

We use rocks in many ways, but getting them out of the ground can be difficult. Explosives are often used to blast rocks out of cliff faces. Here limestone is being blasted from a quarry. Above the quarry face, large machines have been used to drill a line of holes into the ground. The holes are packed with explosives, which are detonated from far away.

LOOKING AT ROCKS

THE best way to learn about rocks and minerals is to look closely at as many different types as you can find. Look at pebbles on the beach and the stones in your garden. You will find that they are not all the same. Collect a specimen of each different rock type and compare them with each other. Give each rock a number to identify it and keep a record of where you found it and what you can see in each piece.

A hand lens will magnify your rocks and help you see details that cannot be easily seen with the naked eye. To find out how many different minerals there are in each of your specimens, look for different colours, shapes and hardness. Testing the properties of minerals, such as hardness, can help to identify what sort of rock it is. Ask an adult to take you to the nearest geological museum to compare it with the specimens there. Why not start your own museum at home or at school?

A closer look
Clean a rock with a stiff brush and water. Stand so that plenty of light shines on the rock and experiment to find the correct distance from the hand lens to the rock.

safety glasses

pencils

hard hat

rucksack

chisel

geological hammer

hand lens

water

gloves

penknife

notebook

camera

field guide

compass

bucket

collecting bags and labels

map, mark on the map the places where you found your best specimens

newspapers, for wrapping your specimens in

Rock collecting
Here is the essential equipment that you will need for collecting rocks. Wear protective clothing and always take an adult with you when you are away from home. Safety glasses will protect your eyes from razor-sharp splinters when hammering rocks. Do not attack cliffs or quarry faces with your hammer but keep to blocks that have fallen well away. Always remember that cliffs can be very dangerous – a hard hat or helmet will protect your head from falling rocks. Do not be greedy when collecting. Rocks and minerals need protecting just as much as wildlife and some sites are protected by law. If you are collecting on a beach with cliffs behind, be careful not to be cut off by the tide.

PROJECT

TESTING FOR HARDNESS

You will need: *several rock samples, bowl of water, nail brush, coin, glass jar, steel file, sandpaper.*

1 Clean some rock samples with water using a nail brush. Scratch the rocks together. On the Mohs scale, a mineral is harder than any minerals it can make scratches on.

2 A fingernail has a hardness of just over 2. Scratch each rock with a fingernail – if it scratches the rock, the minerals of which the rock is made have a hardness of 2 or less.

The hardest natural mineral is diamond, with a hardness of 10. It will scratch all other minerals.

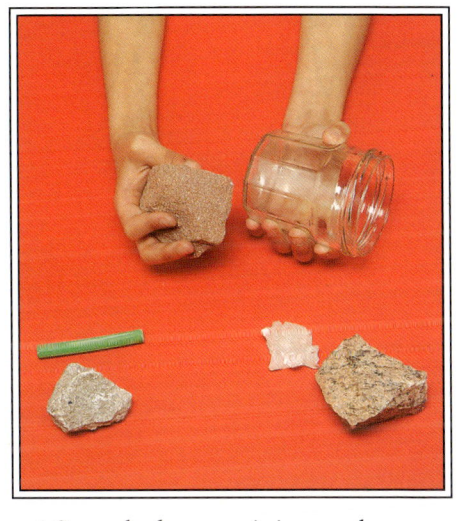

3 Put aside those rocks scratched by a fingernail. Scratch those remaining with a coin. A coin has a hardness of about 3, so minerals it scratches are less than 3.

4 Scratch the remaining rocks on a glass jar. If any scratch the jar, then the minerals they contain must be harder than glass.

THE MOHS SCALE

• Mineral hardness is measured on a scale devised in 1822 by Friedrich Mohs. He listed ten common minerals running from 1, the softest, (talc) to 10, the hardest, (diamond).

5 Put aside any rocks that will not scratch the glass. Try scratching the remainder with a steel file (hardness 7) and finally with a sheet of sandpaper (hardness 8).

WHAT ARE MINERALS?

MINERALS are natural chemical substances that are present in all rocks. Most minerals are solid, but a few are liquid. Some minerals, such as sulphur and gold, are single elements. Others are made up of two or more elements. All rocks are a mixture of minerals. The igneous rock basalt, for example, which makes up most of Earth's oceanic crust, is a mixture of the minerals feldspar and pyroxene. Feldspar itself is a compound of oxygen, silicon and aluminium with various other elements. Silicates are the largest group of rock-forming minerals, all of which include silicon and oxygen. Quartz (the commonest mineral in the Earth's crust) is a silicate. The minerals inside a rock usually form small crystal grains that are locked together to form a hard solid.

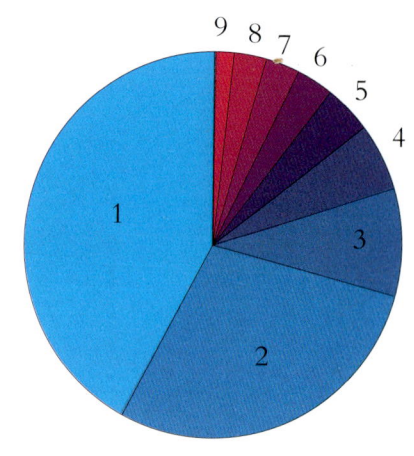

Just eight elements make up nearly all minerals on Earth. Starting with the commonest first, they are oxygen (1), silicon (2), aluminium (3), iron (4), calcium (5), sodium (6), potassium (7) and magnesium (8). All other elements make up (9).

mica

Rock-forming minerals

Granite is one of the most common rocks found in the Earth's land crust. It is made mostly of quartz and feldspar with smaller amounts of mica and hornblende. As molten rock in the Earth's crust cools, the minerals form crystals and interlock with each other. Feldspar crystals are the first to crystallize and may be larger and more perfect in shape than the other minerals, which crystallize later. Feldspar is light (often pink) in colour and quartz is grey and glassy. Mica is dark and silvery, while hornblende is usually jet black. Different granites have different amounts of each mineral, which is why granite varies in colour from grey to reddish-pink.

hornblende

granite

feldspar

quartz

gold

Pure gold

A few minerals occur as single elements. A single element is one that is not combined with any other element. Gold is a good example of a single element. The mineral gold originally comes from hot rocks buried deep underground. Water flows through the hot rocks and dissolves the gold. Sometimes this water moves up towards the Earth's surface, cooling as it does. At lower temperatures the dissolved gold starts to harden and crystallizes into the solid form seen here.

emerald in mass of mica schist

Single element

Diamonds grow under extreme pressure deep in the Earth and are carried to the surface in a rare volcanic rock known as kimberlite. The mineral diamond contains a single element, which is carbon.

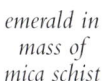

Real versus synthetic

Gemstones such as diamonds and emeralds are rare and expensive. Today the finest quality emeralds are found in the mountains of Colombia and Brazil, in South America.

synthetic emerald

Colourful minerals

Under a microscope, a rock's crystals appear large enough to study. Scientists can identify the minerals by using filters that make polarized light. This gives each mineral its own range of colours.

Man-made crystals

Synthetic crystals can be made to grow in a particular size and shape, for a specific purpose. This is done by subjecting the crystals of more common minerals to carefully-controlled temperature and pressure. Some are used in the electronics industry for making computer chips. Others are manufactured for use in jewellery, such as the synthetic emerald cluster above.

CRYSTALS

pyrite

selenite

quartz

topaz

M OST minerals found in rocks are in crystal form. They are highly prized for their beautiful colours and because they sparkle in the light. Crystals have often been associated with magic – the fortune teller's crystal ball was originally made from very large crystals of quartz. Most precious gemstones, including diamonds and rubies, are crystals.

Igneous rocks are usually made of interlocking crystals that form as hot magma (liquid rock) cools. The largest and best crystals are found in rock features known as veins. Veins are formed when hot, mineralized water rises up through the Earth. As the water cools, crystals form. Crystals may also grow when water on the Earth's surface evaporates. Each mineral variety forms crystals with a characteristic shape.

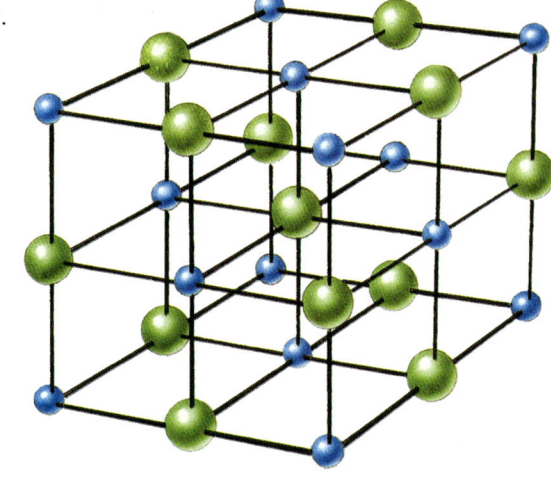

Salt magnified

These grains of ordinary table salt, seen under a powerful microscope, grew when salt water was evaporated. The grains are minute crystals, each containing billions of tiny particles known as atoms. The grains of salt are shaped like tiny cubes because of the way the atoms are arranged inside them. Each variety of mineral forms crystals with an individual shape.

calcite

Crystal faces

Crystals sparkle because their surfaces, or faces, reflect the light. Each individual mineral or group of minerals has faces that are always at the same angles relative to each other. There are seven main groups of crystals based on the arrangements of faces. The mineral crystals shown here illustrate five of the different groups.

Inside crystals

The shape of a crystal is controlled by the way that the atoms inside it are arranged. Imagine that the oranges in these boxes are atoms. In the left box the atoms are stacked in a disorderly way. Atoms that join like this do not produce crystals. Instead, they produce a material called a glass. In a crystal, atoms join together in an orderly way, as in the box on the right.

The crystal lattice

The atoms in a crystal link together to form a three-dimensional framework known as the crystal lattice. This repeats itself in all directions as the crystal grows, giving the crystal its regular shape and controlling the angles between the faces. This picture shows ordinary salt or halite. The green balls represent chlorine atoms, the blue ones are sodium.

reniform, haematite

lamellar, muscovite mica

rosette, gypsum (desert rose)

acicular, pyrite

How does your crystal grow?

A mineral's habit is the shape in which its crystals grow. Different habits form according to the conditions in which crystals grow. Each habit is the result of the crystal lattice framework growing more in some directions than others. A selection of habits is shown here. Each habit has a different name which describes the way it looks. *Acicular* are needle-like crystals, *lamellar* means paper thin, *reniform* are kidney-shaped, *dendritic* are tree-like.

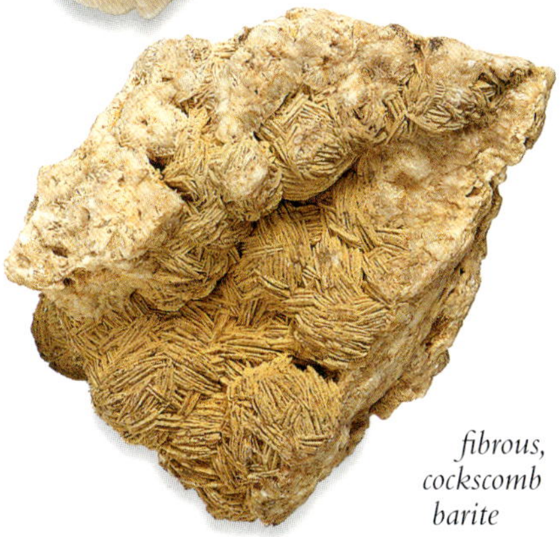

fibrous, cockscomb barite

acicular, gypsum (daisy gypsum)

dendritic, manganese oxide

prismatic, amethyst

Crystal colours

The colour of a crystal in natural light is a useful aid to identifying its mineral. Many types of mineral have characteristic colours, but several occur in a variety of colours. Quartz, for example, can be white, grey, red, purple, pink, yellow, green, brown, black and colourless. Citrine, rock crystal and rose quartz are three types of quartz. Amethyst is purple quartz.

rock crystal

citrine

rose quartz

Crystal twins

Sometimes crystals form so that two (or more) seem to intergrow symmetrically with each other. These are called twin crystals. Aragonite *(above)* often grows twinned crystals.

MAKING CRYSTALS

M OST solid substances, including metals, consist of crystals. To see how crystals form, think what happens when sugar is put into hot water. The sugar dissolves to form a solution. If you take the water away again, the sugar molecules are left behind and join up to reform into crystals. See this happen for yourself by trying the project below. Crystals can also form as a liquid cools. The type of crystals that form will depend on which substances are dissolved in the liquid. In a liquid, the atoms or molecules are loosely joined together. They can move about, which is why a liquid flows. As the liquid solidifies, the molecules do not move around so much and will join together, usually to form a crystal. You can see this if you put a drop of water on to a mirror and leave it in a freezer overnight. Finally, make a simple goniometer a device used to measure the angles between the faces of some objects.

Ice crystals
A drop of water placed on to a dry mirror will spread out a little, then freeze solid in the freezer. Examine the crystals that form with a hand lens.

You will need: water, measuring jug, saucepan, sugar, tablespoon, wooden spoon, glass jar.

GROWING CRYSTALS

1 Ask an adult to heat half a litre of water in a saucepan until it is hot, but not boiling. Using a tablespoon, add sugar to the hot water until it will no longer dissolve.

2 Stir the solution well, then allow it to cool. When it is quite cold, pour the solution from the pan into a glass jar and put it somewhere where it will not be disturbed.

3 After a few days or weeks, the sugar in the solution will gradually begin to form crystals. The longer you leave it, the larger your crystals will grow.

MAKING A GONIOMETER

1 Firmly hold a protractor on a piece of card. Draw carefully around the protractor on to the card using a dark felt-tip pen or soft pencil. Do not move the protractor.

2 With the protractor still in place, mark off 10-degree divisions around the edge. Remove the protractor and then mark the divisions inside the semicircle.

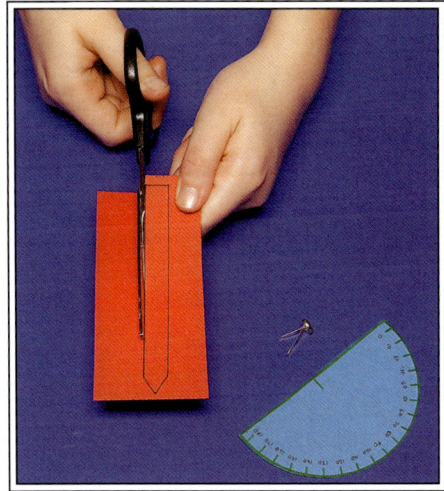

3 Cut out the semicircle. Now cut a thin strip of card about two centimetres longer than the base of your semicircle. Cut one end square and cut the other end into a point.

You will need: *protractor, two pieces of card, felt-tipped pen or pencil, scissors, ruler, paper fastener.*

Measuring the angles

People who study crystals sometimes use a device called a goniometer. It measures the angles between the faces of a crystal. The angle can help to identify a mineral.

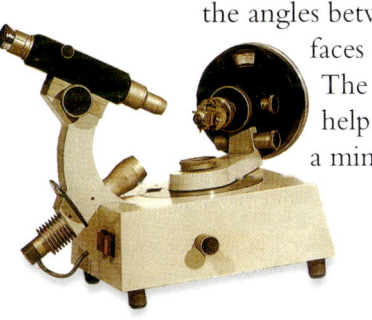

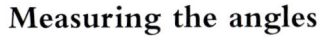

a simple goniometer

4 Make holes in the pointer and semicircle big enough for the paper fastener, and fix them together with the fastener as shown. Flatten out the fastener on the back.

5 Collect some objects with straight sides or faces. Rest the straight face of your semicircle on one face of the object. Move the blunt end of your pointer on to the next face. The other end will point on the scale to the angle between the faces. Real goniometers are more complex and accurate than this, but they measure angles in a similar way to your home-made one and help to identify minerals.

IGNEOUS ROCKS

IGNEOUS rocks start off deep within the Earth as magma (molten rock). The name igneous means "of fire". The magma rises towards the surface where it may erupt as lava from a volcano, or cool and solidify within the Earth's crust. Igneous rocks that extrude, or push out, above ground are called extrusive. Those that solidify underground are called intrusive. Igneous rocks are a mass of interlocking crystals, which makes them very strong and ideal as building stones.

The size of the crystals depends on how quickly the magma cooled. Lavas cool quickly and contain very small crystals. Intrusive rocks cool much more slowly and have much larger crystals. The commonest kind of lava, basalt, makes up most of the Earth's oceanic crust. Granite is a common intrusive igneous rock. It forms huge plugs, up to several kilometres thick and just as wide, in the continental crust. These are called plutons. They are often found under high mountains such as the Alps or the Himalayas.

Microscopic view
Geologists look at rocks under a special kind of microscope that shows the minerals in the rock. This is what an igneous rock called dolerite looks like. Notice the way the crystals fit together with no spaces between. Dolerite is a volcanic lava that formed beneath the ground. It has larger crystals than extrusive basalt.

Fine-grained granite
The magma that makes granite below the ground can also erupt at the surface. This Stone Age axe is made of a lava called rhyolite, and has razor-sharp edges. Its crystals are tiny.

Coarse-grained granite
This sample of granite is typical of igneous rocks. The large crystals give the rock a grainy texture. The crystals are large because they grew slowly as the liquid magma cooled down slowly.

Fine-grained basalt
Basalt is the most common extrusive igneous rock, especially in the oceanic crust. Basalt cools much more quickly than granite, so the crystals are smaller and the rock looks and feels smoother.

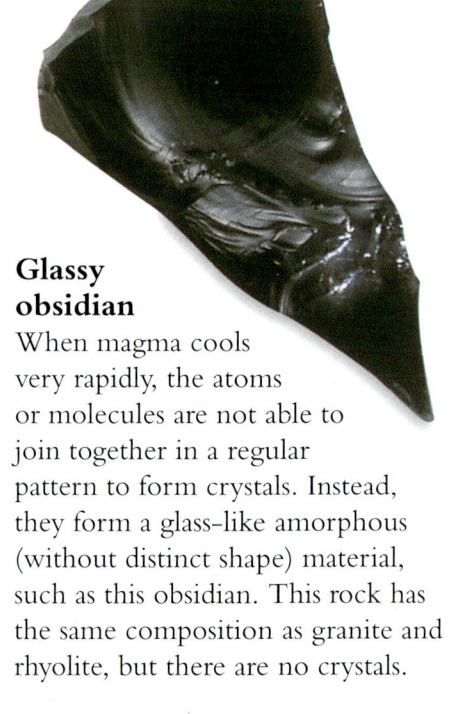

Glassy obsidian
When magma cools very rapidly, the atoms or molecules are not able to join together in a regular pattern to form crystals. Instead, they form a glass-like amorphous (without distinct shape) material, such as this obsidian. This rock has the same composition as granite and rhyolite, but there are no crystals.

Half dome

Millions of years ago, a huge dome of magma intruded under what is now Yosemite National Park in California. It slowly cooled to form granite. Over a period of time, the rocks around it were worn away by glaciers, exposing these dome-shaped hills of granite behind.

Granite tors

On Dartmoor in the southwest of England are the remains of large granite plutons that solidified below a chain of mountains. The mountains were eroded away, but the granite, being hard and weather-resistant, remains to form shapes known as tors. They look like huge boulders stacked on top of each other.

Rivers of liquid basalt

The islands of Hawaii are the exposed tops of huge piles of basalt that is still erupting after many thousands of years. The hot lava glows red in the dark and is capable of flowing for several miles before solidifying. Parts of Scotland and Ireland looked like this about 50 million years ago.

As the lava of the Giant's Causeway cooled, it cracked into interlocking, six-sided columns of basalt rock.

Giant's Causeway

The impressive columns of the Giant's Causeway in Northern Ireland are solid basalt. As lava reached the surface, it flowed into the sea, where it cooled and split into mainly hexagonal (six-sided) columns. The minerals that make up basalt, such as feldspars, pyroxenes and olivine, typically give the rock a dark-gray to black color.

MAKING IGNEOUS ROCKS

THE projects on these pages will show you how igneous rocks can be grainy and made of large crystals, or smooth and glassy. You will be melting sugar and then letting it solidify. Sugar melts at a low enough temperature for you to experiment with it safely at home. To make real magma, you would need to heat pieces of rock up to around 1,000°C until it melted! Even with sugar, the temperature must be high, so ask an adult to help you while you are carrying out these projects. You can also make the sugar mixture into bubbly honeycomb, a form similar to the rock pumice. It is like pumice because hundreds of tiny bubbles are captured inside the hot sugar.

honeycomb (pumice)

toffee (obsidian)

fudge (granite)

Fudge's grainy texture is similar to granite. Like obsidian, glassy toffee cools too rapidly to form crystals. The bubbles in honeycomb are like pumice.

You will need: *sugar, water, saucepan, safety glasses, wooden spoon, milk.*

MAKING CRYSTALLINE ROCK

1 Ask an adult to heat 500g of sugar with a little water in a pan. Continue heating until the mixture turns brown, but not black, then add a dash of milk.

2 Leave the mixture in the pan to cool at room temperature. After an hour, you should see tiny crystal grains in the fudge mixture. Once it is completely cool, feel its texture in your hands.

MAKING GLASS AND BUBBLES

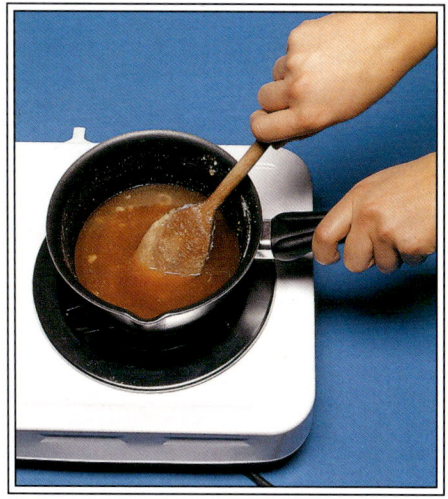

1 Use greaseproof paper to spread the butter over a metal baking tray. Put in the freezer for at least an hour to get cold. Use oven gloves to take the tray from the freezer.

2 Ask an adult to heat about 500g of sugar with a little water in a saucepan. The sugar dissolves in the water, but the water soon evaporates, leaving only sugar.

3 Stir the sugar mixture with a wooden spoon while it is heating. Make sure that the sugar does not burn and turn black. It should be golden brown.

You will need: *greaseproof paper, butter, baking tray, oven gloves, sugar, water, saucepan, wooden spoon, safety glasses, bicarbonate of soda.*

Bicarbonate gives off the gas carbon dioxide to form tiny bubbles.

5 To make honeycomb, stir in a spoonful of bicarbonate of soda in Step 3, just before you pour the sugar on to the tray. This will make tiny bubbles of gas in your "magma".

4 Pour the mixture on to the cool baking tray. After 10 minutes, the glassy and brittle toffee will be cool enough to pick up.

SEDIMENTARY ROCKS

Many of the most familiar rocks that we see around us are sedimentary rocks. Particles of rock, minerals and the shells and bones of sea creatures, settle in layers and then harden into rock over thousands of years. Rock particles form when other rocks are eroded (worn down) by the weather and are carried away by wind, rivers or ice sheets. They become sediments when they are dumped and settle. Sediments may collect in areas such as river deltas, lakes and the sea. Very large particles make conglomerates (large pebbles cemented together), medium-sized ones make sandstones and very fine particles make clays. Some sediments are made entirely of seashells. Others form when water evaporates to form a deposit called evaporite. Rock salt is a sedimentary rock and is used to make table salt.

Sandstone monolith
Uluru (Ayers Rock) is a monolith (single block of stone) in central Australia. It is the remains of a vast sandstone formation that once covered the entire region.

Clay
The particles in clay are too fine even to see with a microscope. Clay absorbs water, which makes it pliable and useful for modeling.

Limestone
This is one of the most common sedimentary rocks. It forms in water and consists mainly of the mineral calcite. Rainwater will dissolve it.

Conglomerate
A conglomerate contains rounded pebbles cemented together by rock made of much smaller particles.

Chalk
Chalk is made from the skeletons of millions of tiny sea creatures. The white cliffs of Dover on the southern coast of England are chalk.

Sandstone
There are many types and colors of sandstone. Each different type is made of tiny grains joined together. The grains are usually quartz.

Red sandstone
The quartz grains in this rock are coated with the mineral hematite (iron oxide) to give the red color. The rock is from an ancient desert.

Old and new

This cliff, on the coast of Dorset, in southern England, is made of layers of hard limestone and soft mudstone. Both rocks were once at the bottom of a shallow tropical sea at a time when dinosaurs roamed on land. The sea was teeming with all kinds of creatures that were buried and became fossils. The cliff is falling (making it dangerous) and will make new sediment as it is broken up by the waves and carried into the sea.

Beach pebbles

To see how the particles in sedimentary rocks form, look at the different sizes of pebbles on a beach. The constant back and forth of the waves grinds the pebbles smaller and smaller. When they settle and the conditions are right, these particles will form sedimentary rock.

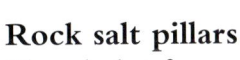

Meeting place

Many sedimentary rocks form at deltas, where a river meets the sea. The river's flow slows right down, so sediment can no longer be carried in the water and is deposited. This picture of a delta on the island of Madagascar was taken from a satellite.

Rock salt pillars

The salt that forms rock salt is a chemical compound called sodium chloride, or halite. Rock salt forms as water evaporates from a salt solution, such as sea water. Here, pillars have formed in the extremely salty water of the Dead Sea, on the border between Israel and Jordan.

The red color of this halite is caused by impurities.

As water dries out, rock salt forms in pillars at the lake's edge.

FACT BOX

• Sedimentary rocks are said to be lithified, which means turned to stone. The word comes from the Greek word *lithos* (stone). The solid part of the Earth (including the crust) is called the lithosphere.

• Large areas of the world are covered with a yellow sedimentary rock called loess. The word comes from an old German word meaning loose. This is because loess consists of tiny dust particles that settled after being blown long distances by the wind.

ROCKS IN LAYERS

SEDIMENTARY rocks form as small particles of rock accumulate at the bottom of seas and lakes, or in deserts. These particles settle to cover large areas and, over thousands or millions of years, new layers of sediment are laid down on top of existing ones.

As a result, most sedimentary rocks form in layers, called strata. The strata that are deepest underground are the oldest, because more recent layers are laid down on top of them. For this reason, sedimentary rock strata can provide valuable clues about the distant history of the Earth.

Once formed, sedimentary rocks may be subject to powerful forces caused by the movement of the Earth's crust. The forces squeeze the strata into folds and crack them. Along some large cracks, known as faults, blocks of rock slide past each other. Both folds and faults are often clearly visible in rock faces.

Strata sandwich

A multi-layer sandwich is like rock strata. The first layer is a slice of bread at the bottom. Each filling is laid on top with more slices of bread. When it is cut through you can see the many different layers.

Folded strata

This rock face shows what happens when parts of the Earth's crust are pushed together in a collision zone. Geologists call downward folds synforms and upward folds antiforms. Folds are found in many sizes from the microscopic to the gigantic. Mountain ranges have very large folds, sometimes with strata turned upside down. Old rocks may have been folded many times.

Layers of rock

Rock faces in cliffs, river valleys and mountains reveal sedimentary rock. Here you can see different layers of the same rock repeated. The layers are not of equal thickness. This suggests that conditions in this region changed many times in the past. Unequal thicknesses are common in layers of sediment found deposited at the mouths of large rivers.

Geological time chart

Era	Period	Million years ago
Cenozoic	Quaternary	
	Holocene (epoch)	0.01
	Pleistocene (epoch)	2
	Tertiary	
	Pliocene (epoch)	5
	Miocene (epoch)	25
	Oligocene (epoch)	38
	Eocene (epoch)	55
	Paleocene (epoch)	65
Mesozoic	Cretaceous	144
	Jurassic	213
	Triassic	248
Paleozoic	Permian	286
	Carboniferous	360
	Devonian	408
	Silurian	438
	Ordovician	505
	Cambrian	590
Pre-Cambrian		4,600

Layers of seashells

In some sedimentary rocks there are layers made mostly of shells. In this picture the rock has split along the shell layer. It is like looking at the filling in a sandwich from above, after the top slice of bread has been taken away.

Rocks and time

To put events into their correct place in time, it is necessary to have a framework or calendar that divides time in a way that is understood by everyone. Geologists have devised their own special calendar which is known as the Geological Timescale. This calendar starts 4,600 million years ago, which is the date of the oldest known rocks people have discovered. In this calendar, there are four major subdivisions, known as Eras. These are subdivided into Periods or Epochs (as shown in the chart above). So if you tell someone that you have found a Miocene fossil, he or she will know how old it is by referring to the calendar.

Clues to the past

Layers of sedimentary rock sometimes look different from one other. The differences are often evidence of climate changes in the past. Other differences might have been caused by the movement of tectonic plates.

Grand Canyon

The bottom strata of the 1.6km-deep Grand Canyon, in Arizona, USA, are more than 2,000 million years old. Those at the top are about 60 million years old.

PROJECT

MAKING SEDIMENTARY ROCKS

To help you understand the processes by which sedimentary rocks are made and how they form distinct layers called strata, you can make your own sedimentary rocks. Different strata of rock are laid down by different types of sediment, so the first project involves making strata of your own, using various things found around the kitchen. The powerful forces that move parts of the Earth's crust often cause strata to fold, fault or just tilt and you can see this, too. In the second project, you can make a copy of a type of sedimentary rock called a conglomerate, in which small pebbles and sand become cemented into a finer material. Conglomerates in nature can be found in areas that were once under water.

The finished jar with its layers imitates real rock strata. Most sediments are laid down flat, but the forces that shape the land may tilt them, as here.

You will need:
large jar, modeling clay, spoon, flour, kidney beans, brown sugar, rice, lentils (or a similar variety of ingredients of different colors and textures).

YOUR OWN STRATA

1 Press one edge of a large jar into a piece of modeling clay, so that the jar sits at an angle. Slowly and carefully spoon a layer of flour about 1 inch thick into the jar.

2 Carefully add layers of kidney beans, brown sugar, rice, lentils and flour, building them up until they almost reach the top of the jar. Try to keep the side of the jar clean.

3 Remove the jar from the clay and stand it upright. The different colored and textured layers are like a section through a sequence of natural sedimentary rocks.

MAKING CONGLOMERATE ROCK

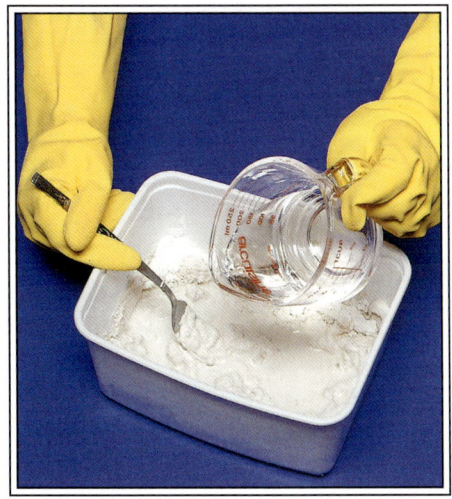

1 Put on a pair of rubber gloves. In a container, mix up some plaster of Paris with water, following the instructions on the package.

2 Mix some small pebbles, sand and soil into the plaster of Paris. Stir the mixture thoroughly to make sure they are all evenly distributed.

3 Let sit for 10 minutes, until the mixture starts to harden, then mold a small lump of it into a ball in your hand.

finished rocks

You will need:
rubber gloves, old plastic container, plaster of Paris, water, fork or spoon, pebbles, sand, soil, paper.

4 Make some more conglomerate rocks in different sizes with different amounts of pebbles. Place the rocks on a piece of paper to harden and dry out completely.

The real thing
Do your rocks look anything like this boulder of real conglomerate rock? The many fragments in this boulder vary a great deal in size and color.

METAMORPHIC ROCKS

THE word metamorphic means "changed" and that is exactly what these rocks are. Metamorphic rocks form when igneous or sedimentary rocks are subjected to high temperatures or are crushed by huge pressures underground. Such forces change the properties and the appearance of the rocks. For example, the sedimentary rock limestone becomes marble, which has a different texture and new minerals that are not found in the original limestone.

There are two types of metamorphism. In contact metamorphism, hot magma heats the surrounding rocks and changes them. In regional metamorphism, deeper rocks are changed when sections of the Earth's crust collide. In the intense heat and pressure of these collision zones the rocks start to melt in some places, new minerals appear and the layers are pushed into strange shapes.

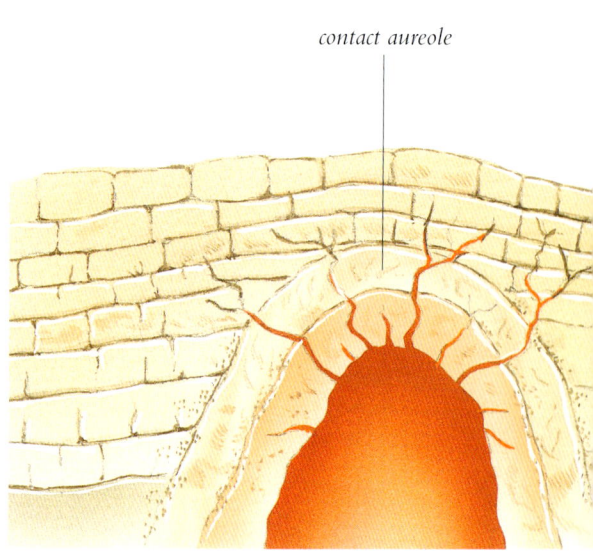

contact aureole

Contact metamorphism
This type of metamorphism occurs when an intrusion of magma bakes the rocks surrounding it. The zone of rock surrounding the magma is called a contact aureole. The magma alters the form and composition of the rocks. The changes are most noticeable close to the intrusion and gradually become less obvious farther away.

Mica crystals in slate give it a shiny, wet appearance.

The minerals form in bands, giving a foliated (layered) appearance.

The mineral olivine gives this marble its green colour. The rest is mostly calcite.

Slate
This rock is formed from mudstone or shale, which are sedimentary rocks containing tiny particles of clay. Slate forms under very high pressure, but at a relatively low temperature. Because of this, fossils from the original rocks often survive but may be squashed out of shape by the pressure.

Gneiss
Under very high temperature and pressure, many igneous or sedimentary rocks can become gneiss (pronounced "nice"). All gneisses are made of layers of minerals. In some, each layer is a different mineral. In others, they are different-size crystals of the same mineral.

Marble
When heat and pressure alter limestone, which is a very common sedimentary rock, marble forms. Impurities in the limestone give marble its many different colours, including red, yellow, brown, blue, grey and green, arranged in veins or patterns.

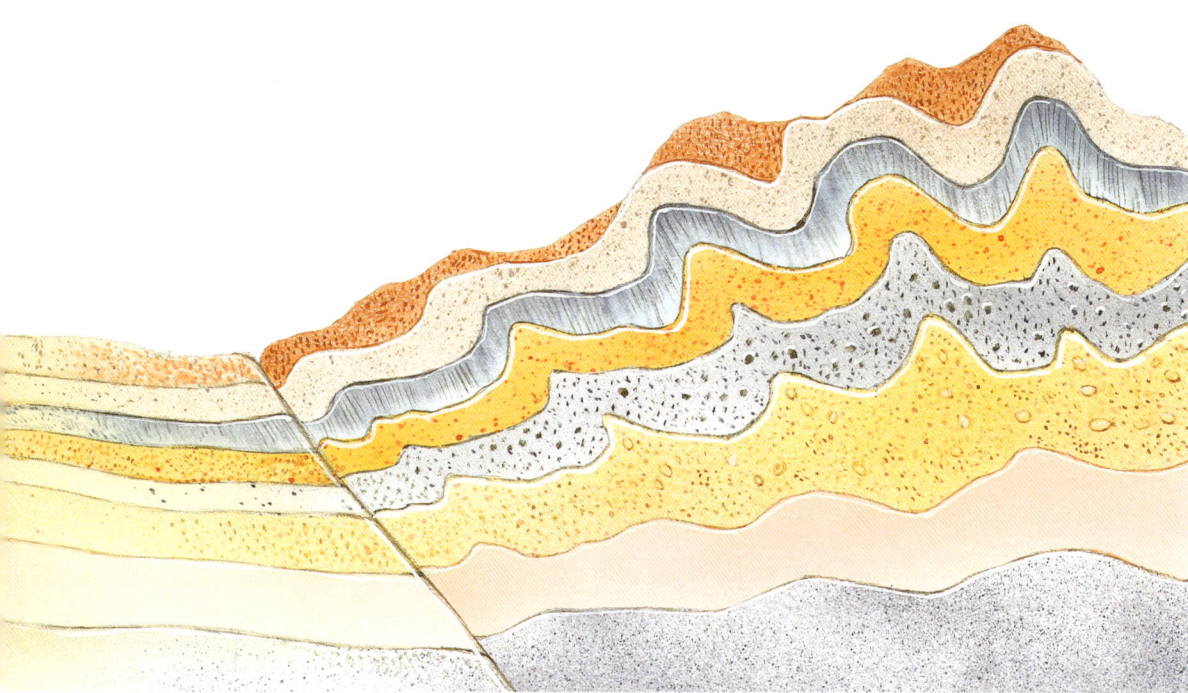

Regional metamorphism

Movements in the Earth's crust can cause rocks to change in many different ways. The nature of the changes depends on the intensity of the pressure and the degree of heat. Slate or schist form at high pressures and low temperatures. For gneiss to form, both temperature and pressure must be very high.

New minerals

Metamorphism causes new minerals to develop in old rocks. The large crystals of garnet in this schist were not in the original rock. The type of minerals that grow depends on the composition of the original rock and the strength of the metamorphism.

Squashed near to melting

When rocks are squeezed and heated, to temperatures close to their melting points, their consistency changes. They are no longer hard, brittle solids that can be cracked – they become plastic (easily moulded). This gneiss was deformed when it was in a plastic state.

Why slate splits

The slate from this Welsh quarry will be used mainly for roof tiles. Slate splits naturally into thin slices along lines within its structure. This is because it has been squeezed. Then, small, lamellar (flattened), lined-up crystals grew in layers at right angles to the direction of the squeeze.

GEOLOGISTS AT WORK

GEOLOGY is the study of the history of the Earth, as revealed the rocks found in the Earth's crust. Scientists who study geology are called geologists. Some geologists specialize in certain branches of geology. For example, paleontologists study fossils, mineralogists specialize in identifying minerals and petrologists study the internal structure and composition of rocks.

Most geologists spend some of their time in the field, collecting samples and measuring various features of the rocks that they can see in outcrops. The rest of their time is spent in laboratories, analyzing the samples they have collected and the measurements they have made, often using a computer. Geologists keep detailed notebooks and make records of everything they discover about the rocks in the area where they are working.

The most important record is a geological map. This shows where different rocks may be found, how old they are and whether they are flat-lying, tilted or folded. This information also gives geologists clues about what is going on underneath the Earth's crust. The different rock types are shown on a map in different colors.

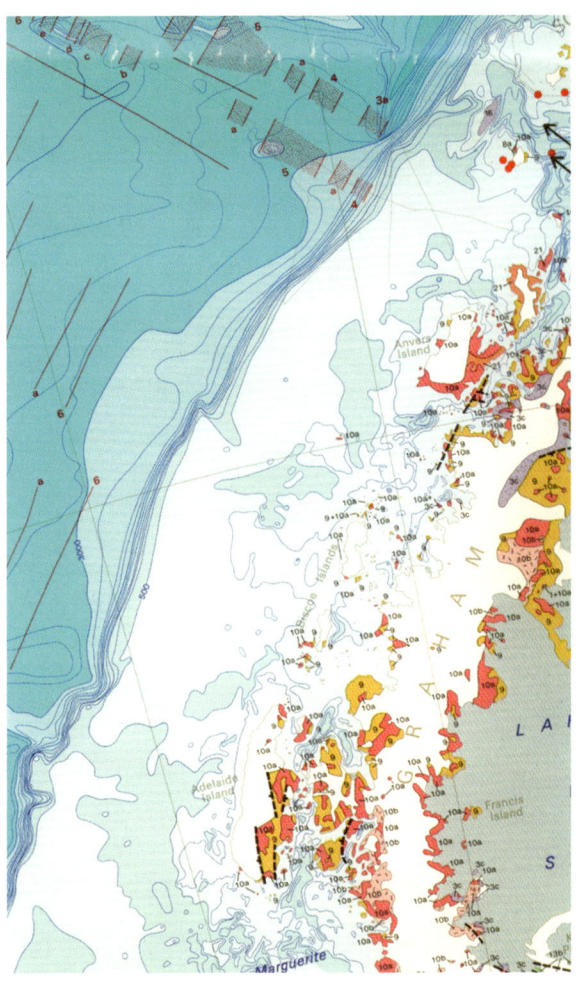

Planning ahead
Geological maps show the types and ages of rocks and how they lie on the land. From such information, geologists can work out where to look for minerals and underground water, and architects decide a suitable place for a building.

Fieldwork
These geologists are sampling water from a volcanic sulfur spring. Most geologists carry out fieldwork like this, collecting various kinds of sample for further study. In this case they will analyze the water to see what minerals have dissolved in it.

Sampling gas
These geologists are collecting samples of gas given off by a volcano. They wear gas masks because some of the gases may be harmful to their health.

Geological hammer

Geologists use a hammer to collect samples of rock. Sometimes they take a photograph of the hammer resting on an outcrop of rock to show the direction and scale of the rock.

FACT BOX

• After rocks have been formed, the radioactive elements in some minerals change into other elements at a steady rate over thousands of years. By measuring the amounts of the new elements, geologists can figure out a rock's age.

• Another way to tell the age of rocks is to look at the fossils they contain. Different plants and animals lived at different times in the past. Identifying fossils can tell you where in geological time a rock fits.

Microscopic photograph

A micrograph is a photograph taken through a microscope. This picture shows the grains in sandstone greatly magnified. Grain size and shape can help to identify sandstones.

Rocks in close-up

Slices of rock are examined under a petrological microscope. This has special filters to polarize light. The slices are cut thinly so that light passes through them. The appearance of minerals in polarized light helps geologists to identify them.

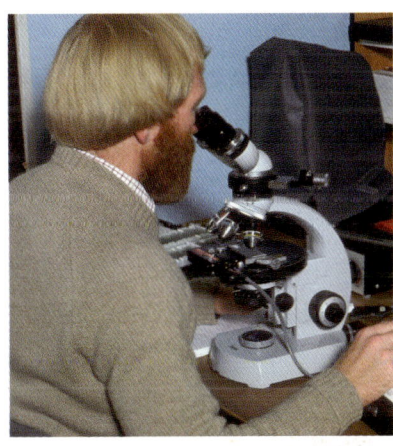

Polarized light micrograph

This micrograph of a thin section of a metamorphic schist was made using a polarizing microscope. It shows the crystals of different minerals that make up the rock. The micrograph helps to identify the rock's origin and the temperature and pressure at which it formed.

Density

These blocks are the same size and shape, but do not weigh the same. The materials they are made of have different densities. Density is used to identify minerals because samples of the same mineral will have the same density.

IDENTIFYING MINERALS

GEOLOGISTS use many different methods to identify the minerals that make up rocks. Each mineral possesses a unique set of identifying properties. Geologists use several tests to identify minerals, such as hardness (how easily a mineral scratches) and specific gravity (comparing a mineral's density to the density of water). They also look at streak (the colour of a mineral's powder), lustre (the way light reflects off the surface), transparency (whether light can pass through or not) and colour (some minerals have a distinctive colour in natural light). Dropping acid on to a sample to see if gas is given off is a simple test that can be carried out in the field. Try these simple versions of two tests that geologists use. They will help you identify some samples that you have collected. Firstly, rubbing a rock on to the back of a tile leaves a streak mark – the colour of the streak can reveal the minerals that are present. Then you can calculate the specific gravity of a sample.

Acid test
Drop a rock into vinegar. If gas bubbles form, then it contains minerals called carbonates (such as calcite).

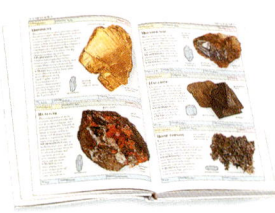

You will need:
white tile, several samples of different rocks or minerals, field guide.

STREAK TEST

1 Place a tile face down, so that the rough side is facing upwards. Choose one of your samples and rub it against the tile. You should see a streak of colour appear on the tile.

2 Make streaks using the other samples and compare the colours. Rocks made of several minerals may leave several coloured streaks.

SPECIFIC GRAVITY TEST

1 Choose a rock and weigh it as accurately as you can to find its mass (weight). The figure should be in grams. Make a note of the mass.

2 Fill a clear measuring jug to the 200ml mark with water. Now carefully place the first rock sample into the water.

3 Look carefully at the scale on the jug to read off the new water level. Make a note of the level of the water in your notebook.

You will need:
mineral or rock samples, accurate weighing scales, notebook, pen or pencil, measuring jug, water.

The mass of a sample divided by its volume gives you its density, or specific gravity.

4 Take the figure you wrote down in Step 3 and subtract 200. This is the sample's volume in millilitres. Now divide the mass (weight) by the volume to find the density. You can use a calculator to do this sum if you wish.

beryl

pyrite

More dense

If a mineral has a specific gravity (SG) of 5, it is five times as dense as water. Pyrite has an SG of 5 and beryl has an SG of 2.6. The atoms in pyrite are more closely packed together, making it denser.

READING THE ROCKS

GEOLOGISTS are interested in what has happened on planet Earth from the time it appeared over 4,000 million years ago to the present day. By reading clues in the rocks they can piece together how the climate changed in the past, how continents moved and how oceans and ice-sheets appeared and disappeared. No single geologist can discover all of this on his or her own. Each individual adds clues that are shared and considered by others in an endless process of detective work.

The first clues come from a careful study of exposed outcrops in the field. What minerals can be seen, and are they crystals? Are the rocks in layers? Are there fossils in them? Are the layers flat or tilted, or bent into shapes like waves? These are just a few of the questions that geologists ask.

More clues come from samples tested in the laboratory. What elements are in the samples? Are the rocks magnetic? How old are they? What kind of fossils do they contain? Each fragment of information is written down or stored on a computer. A store of knowledge about rocks is gradually assembled for geologists, so that they can draw conclusions about Earth's history.

When continents collide

An outcrop of gneiss, like the one above at the surface of the Earth, would tell geologists that the gneiss was once buried beneath high mountains like those on the right. Mountain ranges often develop when moving sections of the Earth's crust have collided, subjecting the rocks to the kind of strong squeezing and heating that creates gneiss. In time, the mountains are eroded away and the gneiss is exposed at the surface. The mountains shown on the right are the Alps, the result of Africa colliding with Europe.

Tropical sea

Limestones like this (left) begin in the sea. Sea creatures play an important part in their development. If you could travel back in time you would find the sea full of strange fish, shells and corals – similar to those in the picture on the right. When these kinds of creatures died, fragments of their shells were then consolidated into limestone.

Strong water

The large, rounded pebbles in this conglomerate outcrop (left) indicate that it came from water with currents strong enough to carry the pebbles. Such sediment could have started on a shoreline like the one on the right, or in a large, fast-flowing river.

Sedimentary basins

On the left, layers of hard sandstone alternate with thinner layers of soft shale, a rock rich in clay. Such sediments were washed into a shallow sea (right) to make what is called a "sedimentary basin". The changes in layers of sediment are caused by changes in the slope of the seabed, the position of the shoreline and the depth of the water.

SEEING INSIDE THE EARTH

How do geologists find out what is deep inside the Earth? Geologists can take measurements at rocky outcrops on land, which provide information about the rocks below. However, only a small proportion of the surface rock is exposed and accessible. Alternatively, geologists may probe down into the Earth. Information obtained from holes made during mining, or from natural caves in limestone areas, shows that the temperature of the Earth increases with depth. Smaller and deeper holes can be made by drilling. Modern drilling machines operate on land or at sea and can collect rock samples from several kilometres down. They are also able to drill sideways from the bottom of a vertical hole, and this increases the amount of information obtained about rocks at deep levels.

Another way of seeing inside the Earth is provided by geophysics, a sister science to geology. Geophysics is concerned with measuring things in rocks such as magnetism, gravity, radioactivity and the way in which rocks conduct electricity and sound. Three-dimensional maps can be made from measurements that show ups and downs of land. These maps show the distribution and formation of underground rocks.

Vibrating Earth

Earthquakes have helped geologists to understand what the Earth is made of. When they occur it is as if the Earth has been hit by an enormous hammer. The Earth vibrates like a ringing bell. Waves of vibration move right through the Earth. They may be detected on the surface at places far away from the earthquake – even on the opposite side of the Earth. Very sensitive instruments called seismographs are used to measure the vibrations of an earthquake. These have been placed all over the world and are continuously switched on, ready to detect the next earthquake wherever it starts.

Earthquake waves travel at different speeds in different materials and will arrive at the seismographs at different times. All of the seismograph records ever taken show that the core of the Earth is mostly liquid.

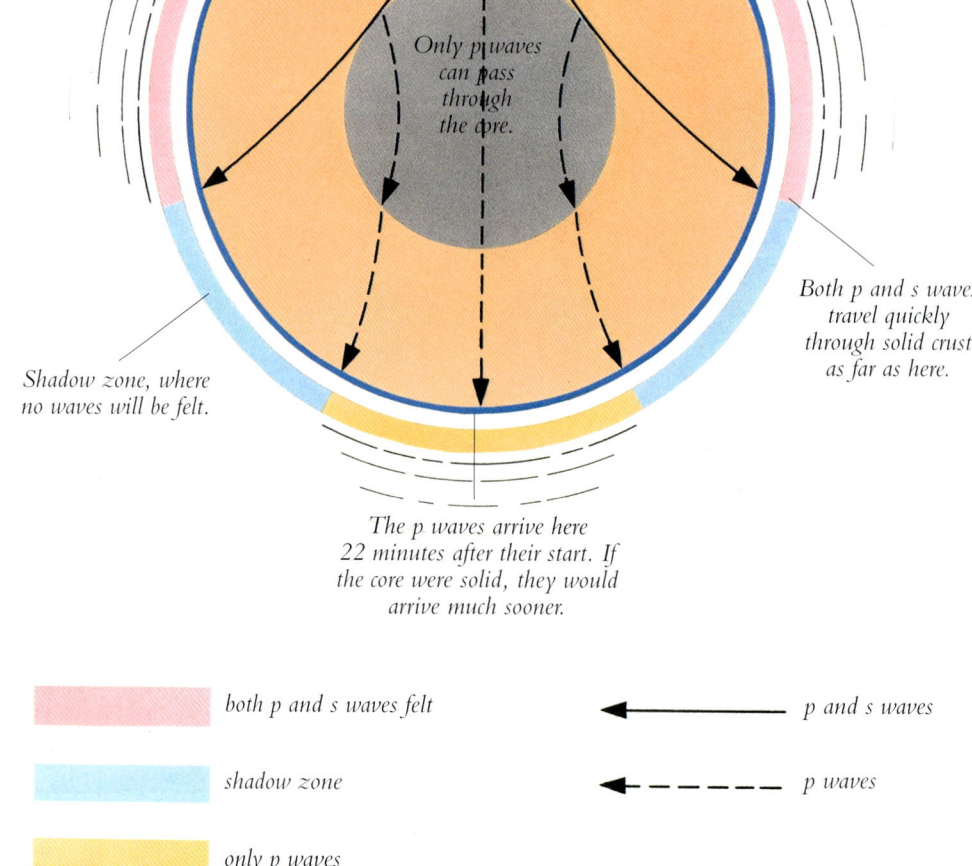

Earthquake starts here and makes vibrations of two kinds – p waves and s waves. The Earth vibrates like a bell as the waves move outwards in all directions.

Only p waves can pass through the core.

Both p and s waves travel quickly through solid crust as far as here.

Shadow zone, where no waves will be felt.

The p waves arrive here 22 minutes after their start. If the core were solid, they would arrive much sooner.

both p and s waves felt

shadow zone

only p waves

p and s waves

p waves

Reflecting sound waves

Bats avoid flying into objects by detecting sound that is reflected from them. Something similar is used to look inside the Earth in seismic reflection surveys. These are commonly done from ships specially designed for the purpose. Loud bangs are made every few minutes by a compressed–air gun towed behind the ship. This sends sound waves down through the water into the rock below. Whenever the sound waves reach a different rock layer, they are reflected back to the surface. The reflections are picked up by a long string of underwater microphones, called hydrophones, also towed behind the ship. Poweful computers analyse the information to read the rocks beneath the seabed.

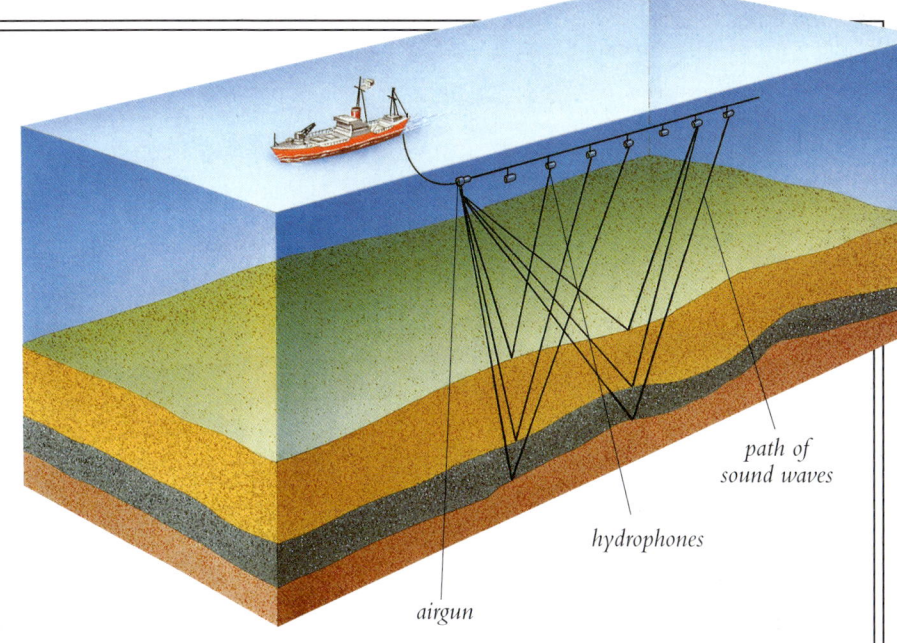

path of sound waves

hydrophones

airgun

Seismic profile

One of the first things to be produced in a seismic reflection survey is a computerized drawing like the one on the left, showing the reflecting layers. From this, a geologist tries to work out what rocks are present.

Drilling

Geopyhsical surveys provide many clues about what lies beneath the ground but they cannot identify rocks. Eventually holes have to be drilled and pieces of rock collected. The holes are made by turning a drill bit, studded with diamonds and attached to the end of a long string of steel pipes. More pipes are added as the bit gets deeper. The samples come to the surface either as broken chips or as solid cylinders of rock, known as core. These men are extracting core from a hole drilled on land.

Handling the data

A modern geophysical survey requires lots of computing power to process information about the inner Earth. This is the computer room in a modern survey ship making a seismic reflection survey. Computers control the airguns on the ship and process the information from each of the geophones.

FINDING BURIED MINERALS

I F we are to use minerals we must first find them. This is the job of a special group of geologists who are experts in mineral exploration. The task is fairly easy if the target mineral lies at the surface of the Earth but most surface minerals have already been found. Nowadays, the search is for minerals that are deeply buried and, in some cases, far beneath the seabed. Firstly the geologist selects a promising area, using his knowledge and experience. The area is then surveyed using a variety of methods, all of which depend on the fact that different minerals have different physical properties. For example, rocks rich in iron are magnetic and may be found using an instrument called a magnetometer which measures magnetism. Other instruments are used to measure things such as gravity, radioactivity or other things which vary from one mineral to another. In this project you will discover how a simple magnetic survey is done using an ordinary pocket compass instead of a magnetometer.

Using magnetism

This project will show you the principle of the use of magnetism in the search for buried minerals. The Earth itself is a giant magnet. The strength and direction of the magnetism in different parts of the Earth is well known to geologists. In places where the rocks contain a lot of iron, the "normal" magnetism will be changed to a stronger magnetism.

haematite (iron ore)

You will need: *plastic tray, clean sand, small magnet, ruler, felt-tipped pens, scissors, stiff card, adhesive putty, small compass.*

USING MAGNETISM

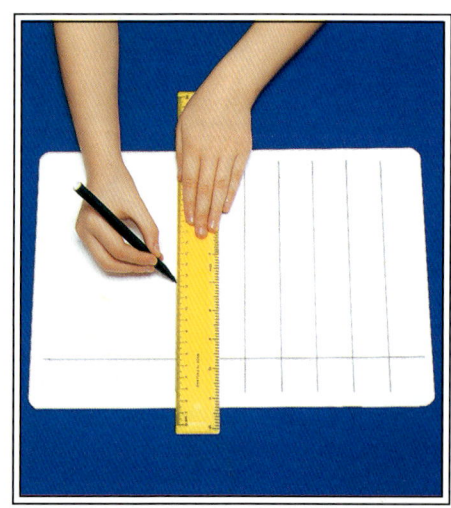

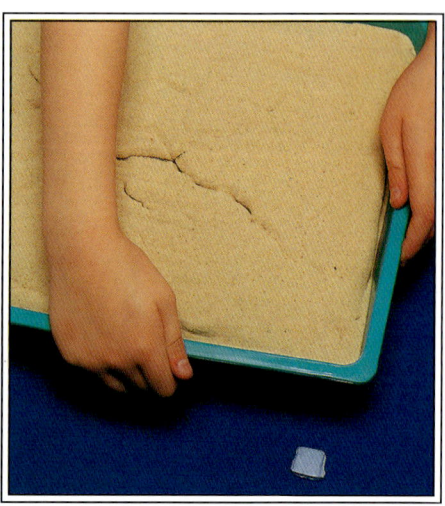

1 Fill a plastic tray nearly to the top with clean sand and ask a friend to bury a small magnet in it while you are not looking and to level the sand until it is smooth.

2 Measure the tray and cut a piece of card to fit inside it. Draw a grid of squares on the card. The lines should be about 30mm apart and parallel to the edges of the card.

3 Using adhesive putty, firmly fix the tray to the table. Make sure that there are no strongly magnetic items, such as radios, nearby or under the table.

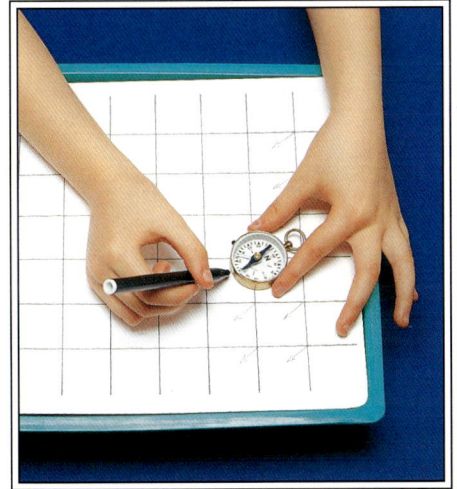

4 Place the compass on one line intersection. When the needle stops moving, hold it down with a finger and make a small dot on the card close to the needle's north end.

5 Take the compass away and draw a line from the dot to the intersection where the compass was sitting. Put a small arrowhead on the north end. Repeat steps 4 and 5.

6 Each intersection will now have an arrow coming from it. Most of them point in one direction. This is magnetic north. Put a red dot on the places where they point differently.

7 The dots mark the places where the compass needle was deflected from magnetic north by the buried magnet. They should be grouped together around the same area. Draw a line enclosing each of the red dots.

You should have found your "iron ore" buried in the sand.

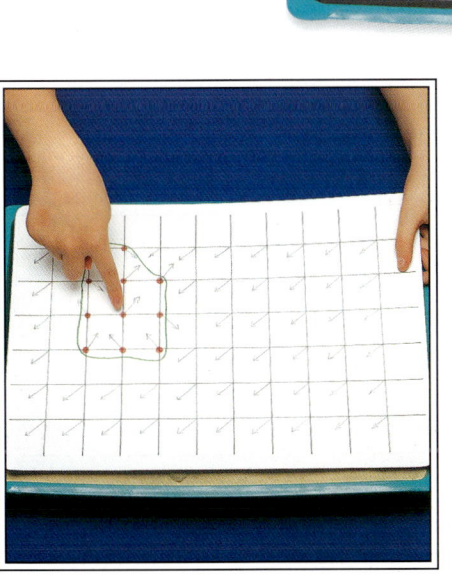

8 Holding your finger above the enclosed area, take the card away and leave your finger poised above the place where the magnet lies buried. Dig in to the sand below.

LIMESTONE LANDSCAPES

calcite

flint nodule

The main mineral in limestone is calcite, which dissolves easily. Lumps or layers of flint or chert (forms of quartz) are often found in limestones.

LIMESTONE is a very common sedimentary rock, made largely of a mineral called calcite, which contains the element calcium. Limestone comes in many forms, chalk being one of them. The dazzling white chalk cliffs of southern England are an example of limestone.

Limestone forms landscapes unlike others, because calcite dissolves quite easily in rainwater. This causes gaping holes to appear at the surface, which may suddenly swallow rivers and divert them underground. Somewhere downstream, the river will usually come out at the surface again.

Underground, many limestones are riddled with caves formed by the action of the slightly acidic water. Caves may form huge interconnected systems, some of which are still unexplored. Where water drips from rocks inside caves, dissolved calcite is deposited (left behind), forming pillars called stalactites and stalagmites. A limestone rock called travertine forms in the same way, leaving a thick coating rather like the fur that forms in a kettle. An extreme limestone landscape called karst occurs where rainfall is fairly high. This is characterized by steep-sided limestone pinnacles separated by deep gullies. China has particularly spectacular karst scenery.

Karst scenery
These pinnacles are in Australia's Blue Mountains. Limestone areas create spectacular landscapes, called karst. Rainwater runs through cracks in the limestone to form underground caves and large holes called sink holes. Where the strata are tilted, deep cracks create pinnacles.

Limestone cones
Cones of limestone rock rise from beside the Li Jiang river near Guilin, China. The strange and beautiful towers were formed by intense downward erosion by rain and river water. The rainfall in this region is very high.

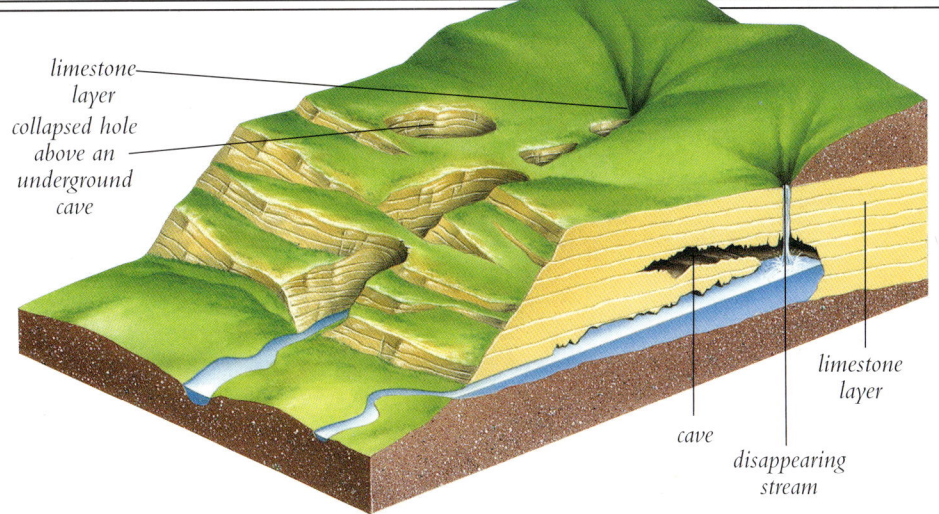

limestone layer

collapsed hole above an underground cave

cave

disappearing stream

limestone layer

Travertine

This famous landscape is the Pamukkale Falls in Turkey. Over thousands of years water from hot springs in limestone regions has deposited travertine (which is of the same composition as limestone) in beautifully shaped terraces. In some places travertine is quarried and used as a decorative building stone.

Limestone, the rock full of holes

Rainwater contains acid which dissolves limestone. As rainwater flows over the surface and through cracks in limestone areas, it slowly dissolves the rock. It leaves behind holes of all shapes and sizes, including potholes and caves. The holes have attracted many people who love to explore them. They are called speleologists (from *spelaion*, the Greek word for cave).

FACT BOX

• Karst scenery covers as much as 15 per cent of the Earth's land area. The word karst was taken from the name of a region on the Dalmatian coast of Croatia, on the Adriatic Sea.

• In regions where the main type of rock is limestone, calcite dissolves into the water supply, forming hard water. In hot water pipes and kettles, some of the mineral is deposited to form brown, furry crystals.

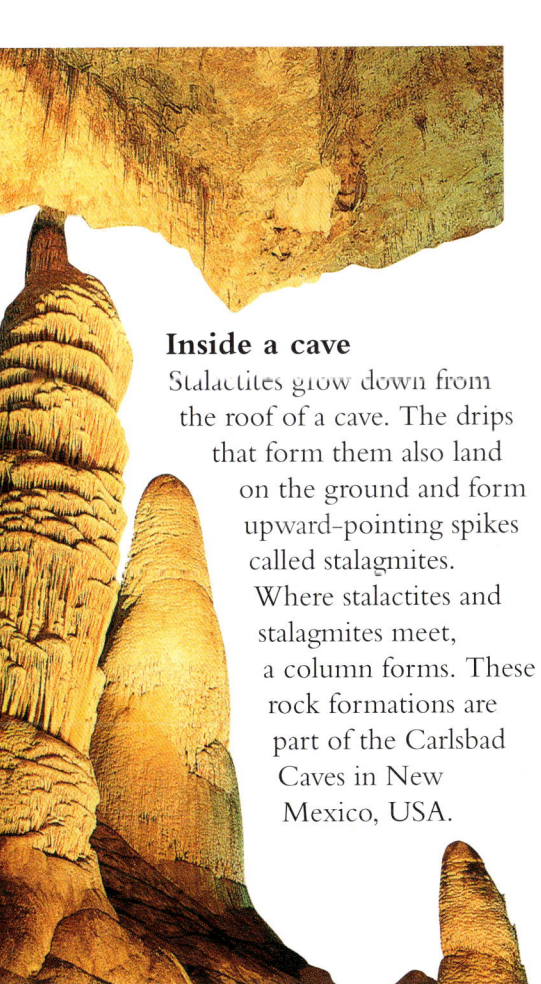

Inside a cave

Stalactites grow down from the roof of a cave. The drips that form them also land on the ground and form upward-pointing spikes called stalagmites. Where stalactites and stalagmites meet, a column forms. These rock formations are part of the Carlsbad Caves in New Mexico, USA.

Limestone pavement

In many limestone regions, pavements with deeply furrowed surfaces are formed. The furrows, known in England as grikes, are made by water seeping through cracks and dissolving the limestone. The blocks of limestone formed by the furrows are called clints.

ROCK, WEATHER AND SOIL

SOIL is another stage in Nature's recycling programme in which one kind of rock is slowly changed into another. Soil is formed from a mixture of mineral grains, pieces of rock, and decayed vegetable matter such as leaves and plants. It forms, by a process known as weathering, when surface rocks are broken down by plant roots and dissolved by water. The action of burrowing animals and insects further helps the weathering process. Some minerals dissolve quickly. Others, such as quartz, are not dissolved but stay behind in the soil as stones.

The soil itself gradually erodes. Particles in it are blown or washed away. There are many different kinds of soil, whose characteristics depend on climate and the type of rock from which they are formed. In hot, wet climates the soils are bright red and thick. In dry or very cold climates the soils are thin or completely absent.

Black sand

Most sand grains pass through soil at some stage in their recycling process. The sand on this beach is made mainly of the black mineral magnetite. It is formed when basalt, a vocanic lava, is broken down through weathering. It then moves in rivers to the sea where it makes black beach sand.

Peat bogs

The amount of plant material in soil can be very high. Peat soil is made up mostly of dead moss.

Sand dunes

In dry climates where deserts form, sand is blown by the wind to form hills called dunes. These red dunes are part of the Namib Desert in south-west Africa. The quartz grains in the sand were once part of the soil that covered the region when the climate was wetter.

Sea mists provide enough moisture for some plants to survive.

Barkhans are crescent-shaped dunes that are always moving.

The wind creates beach-like ripples on the desert floor.

Soil in layers

Soil occurs in layers, known as horizons. There are four main horizons, called A, B, C and R. The A horizon (also called topsoil) is a layer of fine particles that supports the roots of plants and trees. In the B and C layers beneath the topsoil, the soil particles become larger. The R horizon is partly solid rock.

oak seedling

As trees grow, their roots help break down rock into soil. The roots work into cracks, splitting a rock apart.

Dead leaves

As dead leaves and branches rot they release nutrients into the soil. Trees and other plants need these nutrients to grow.

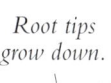

Root tips grow down.

Beach sand

These grains of sand have been enlarged under a microscope. The sand would have originated in soil – as small particles of rock. Over time, the grains have been transported to the sea.

Graded beds

Sand is moved along by flowing water. Where the current slows, sand and rock particles are deposited. This builds up layers of rock called a graded bed. Movements in the crust can tilt the bed at different angles.

River sand

The particles found in river sand, often of quartz, are generally bigger than in desert or beach sand. They are also more angular in shape. The particles come to rest either on the flood plain of a river or at the river mouth in the delta, where they form gritstone, a coarse type of sandstone.

WHAT IS SOIL MADE OF?

SAND and soil are made of millions of very small particles. Sand is formed from many types of rock, by a process called attrition (grinding down). This usually happens after the grains have been released from a soil that formed in a different place at a different time. Desert sand forms by attrition, as wind–blown mineral grains rub against each other. You can see how attrition forms small particles simply by shaking some sugar cubes together in a glass jar.

Soil is a mixture of particles of minerals, along with dead plant and animal matter. In the first project, a sample of soil is examined, using a sieve to separate particles of different sizes. In the second project, you can find out how graded beds of sediment form in rivers, lakes and seas, as first large and then finer particles of sediment are deposited.

Sugar shaker
Shake some sugar cubes in a jar. The cubes knock together as you shake. After a while you will see many tiny grains of sugar. A similar process occurs in desert sands and on beaches as mineral grains knock against each other and become smaller.

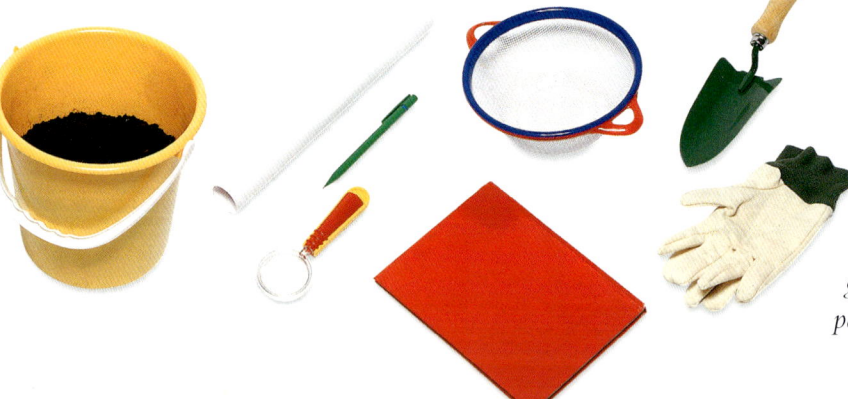

You will need:
gloves, trowel, soil, sieve, paper, hand lens, notebook, pen or pencil.

WHAT IS IN SOIL?

1 Put on the gloves and place a trowel full of soil into the sieve. Shake the sieve over a piece of white paper for a minute or so.

2 Tap the side of the sieve gently to help separate the different parts of the soil. Are there bits that will not go through the sieve?

3 Use a hand lens to examine the soil particles that fall on to the paper. Are there any small creatures or mineral grains? Note what you see.

BIG OR SMALL?

1 Using scissors, cut off the top of a large, clear plastic bottle. Throw away the top part.

2 Put small stones or gravel, earth and sand into the bottom of the bottle. Add water nearly to the top.

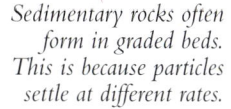
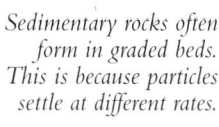

You will need: *scissors, large clear plastic bottle, small stones or gravel, earth, sand, wooden spoon, water.*

3 Mix the stones, gravel, earth, sand and water vigorously together. In a river, rock particles are mixed together and carried along by the moving water.

Sedimentary rocks often form in graded beds. This is because particles settle at different rates.

floating humus and plant fragments

water made cloudy by very fine particles of clay

settled mineral particles

Humus

An important constituent of soil is humus. This is produced by animals called decomposers, such as worms and woodlice. These animals eat dead plant and animal matter, including leaves. As the matter passes through their bodies, it is broken down in their digestive systems.

4 Leave the mixture to settle. You should find that the particles settle in different layers, with the heaviest particles at the bottom and the lightest on top.

PRESERVED IN STONE

coccosphere

radiolaria

foraminifera

THE remains of some organisms (plants and animals) that died long ago can be seen in sedimentary rock as fossils. After an organism dies, it may become buried in sediments. Slowly, over thousands of years, the sediments compact together to form sedimentary rock. The organic remains of the plants or animals disintegrate, but their shape or outline may remain. The hard bits of animals, such as bones and teeth, are preserved by minerals in the rock. Minerals can also replace and preserve the shape of the stem, leaves and flowers of a plant.

The study of fossils, called paleontology, tells us much about how life evolved, both in the sea and on the land. Fossils give clues to the type of environment in which an organism lived and can also help to date rocks. The fossil substance amber was formed from the sticky, sugary sap of trees similar to conifers, that died millions of years ago. The sap slowly hardened and became like stone. Insects attracted to the sweet sap sometimes became trapped in it, died there and so were preserved inside the amber.

Microfossils

Just as there are living organisms too small to see without a microscope (micro-organisms), there are also microfossils. These fossils are tiny marine organisms that lived during the Cretaceous period (about 65 to 144 million years ago). Millions of their remains are found in the sedimentary rock, chalk.

Early animals

Some fossils are the remains of animals that are now extinct. This trilobite, which is 600 million years old, is a distant relation of modern lobsters. Trilobites lived in ancient seas with all kinds of other creatures that are also now extinct.

HOW FOSSILS ARE FORMED

An animal or plant dies. Its body falls on to the sand at the bottom of the ocean or into mud on land. If it is buried quickly, then the body is protected from being eaten.

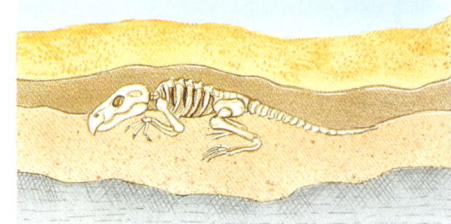

The soft parts of the body rot away, but the bones and teeth remain. After a long time the hard parts are replaced by minerals – usually calcite but sometimes pyrite or quartz.

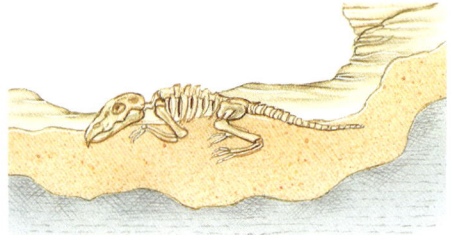

After millions of years the rocks in which the fossils formed are eroded and exposed again. Some fossils look as fresh now as the day when the plant or animal was first buried.

ammonite

amber

Types of fossils

Five common fossils are shown here. Ammonites were hard-shelled sea creatures that lived between 60 and 400 million years ago. Fossils from sea creatures, such as shark's teeth, are often found, because their bones cannot decay completely underwater. The leaf imprint formed in mudstone around 250 million years ago and fern-like fossils are often found in coal. Amber is the fossilized sap of 60 million-year-old trees.

shark's tooth

leaf

fern

FACT BOX

• Perhaps the most interesting fossilized animals are the dinosaurs, which lived between 65 and 245 million years ago. When their shapes are perfectly preserved, expert paleontologists can reconstruct the complete skeleton of the animal.

• Ammonites evolved (developed) rapidly and lived in many parts of the prehistoric world. Because geologists know how ammonites changed, they use the fossils to determine the ages of the rocks in which they are found.

Fossilized crab

This crab lived in the ocean around 150 million years ago. Marine limestones and mudstones contain the best-preserved fossils. Some limestones are made entirely of fossil shells.

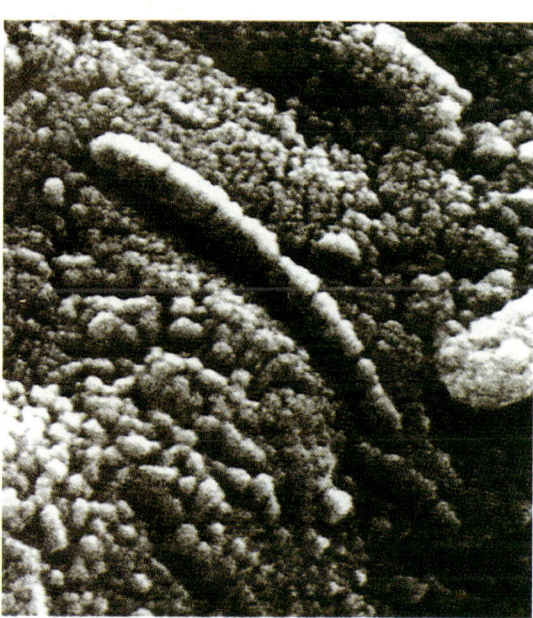

Martian fossils?

In 1996, scientists discovered what looked like tiny fossilized creatures in a rock that had originated on Mars. Everyone was excited about the possible proof that life had once existed on the planet, but it turned out that the marks were probably hardened mineral traces.

MAKING FOSSILS

THESE projects will help you to understand how two types of fossil came to exist. One type forms when a dead plant or animal leaves a space in the sedimentary rocks that settled around them. This is usually how the soft parts of an animal, or a delicate leaf, are preserved before they decay. The space in the rock is an imprint of the dead plant or animal. You can make a fossil of this kind using a shell, in the first project. In this case the shell does not decay – you simply remove it from the plaster. In another kind of fossil, the spaces are formed when the decaying parts of an animal's body or skeleton are filled with minerals. This gives a solid fossil that is a copy of the original body part. Make this kind of fossil in the second project.

These are the finished results of the two projects. While you are making them, try to imagine how rocks form around real fossils. They are imprints of organisms that fell into mud millions of years ago.

These are a good example of dinosaur tracks. They were found in Cameroon, Africa.

You will need: *safety glasses, plastic tub, plaster of Paris, water, fork, strip of paper, paper clip, modelling clay, shell, wooden board, hammer, chisel.*

MAKING A FOSSIL IMPRINT

1 In a plastic tub, mix up the plaster of Paris with water. Follow the instructions on the packet. Make sure the mixture is fairly firm and not too runny.

2 Make a collar out of a strip of paper fixed with the paper clip. Using modelling clay, make a base to fit inside the collar. Press in the shell. Surround the shell with plaster.

3 Leave your plaster rock to dry for at least half an hour. Crack open the rock and remove the shell. You will then see the imprint left behind after the shell has gone.

MAKING A SOLID FOSSIL

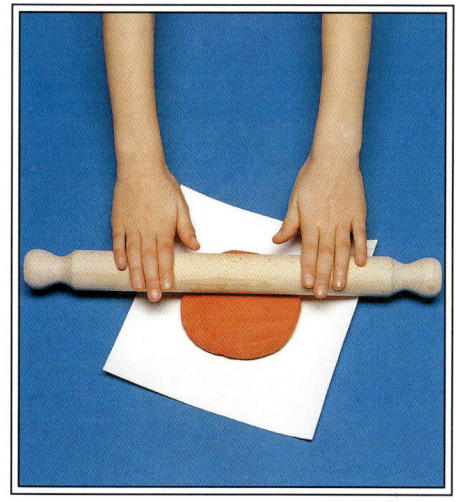

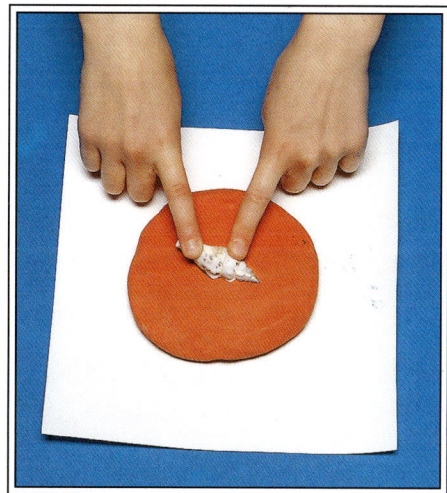

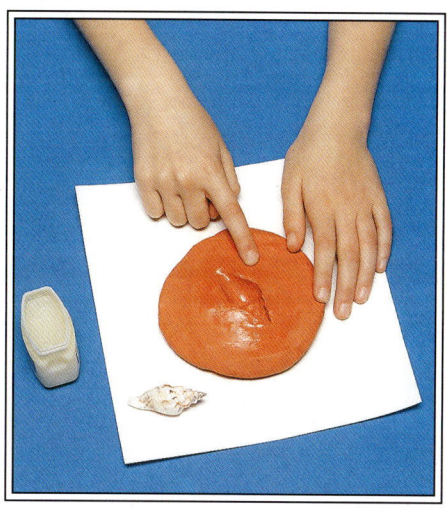

1 Put down a spare piece of paper to protect your work surface. Roll out a flat circle of modelling clay, about 2cm thick.

2 Press your shell or other object deep into the clay to leave a clear impression. Do not press it all the way to the paper at the bottom.

3 Remove the shell and lightly rub some petroleum jelly over the shell mould. This will help you to remove the plaster fossil later.

You will need: *spare paper, rolling pin, modelling clay, shell, petroleum jelly, paper clip, strip of paper, glass, plaster of Paris, water, fork.*

These jewel-like ammonite fossils were found in England at Lyme Regis, Dorset, a source of many different fossils.

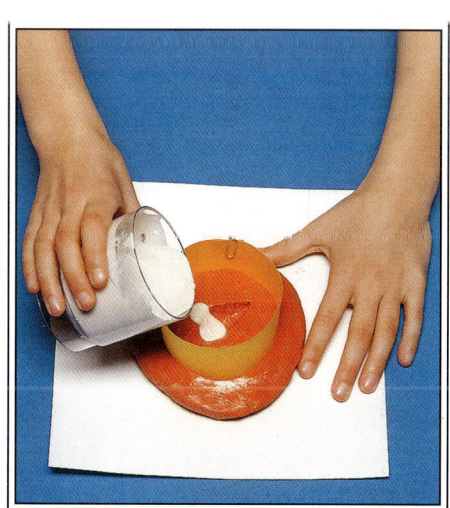

4 With the paper clip, fix the paper strip into a collar for the mould. Mix up some plaster of Paris, pour it in and leave to set for half an hour.

5 Now carefully remove the solid plaster from your mould. In order not to damage them, paleontologists remove fossilized bones or teeth very carefully from rock or earth by cleaning them. They will do some cleaning in the field and the final cleaning back in the laboratory.

USEFUL ROCKS

Aʟʟ kinds of things are made from rocks and minerals. Early humans used hard rocks, such as flint, for stone tools. If you look about your house now, you will see many things that were once rocks or minerals. Bricks and roof tiles are made of clay. The cement that joins the bricks together is made from the sedimentary rock limestone. The plaster on the walls may be from a soft, powdery rock called gypsum. The glass in the windows comes mainly from quartz sand. China and pottery are made of clay. Petrol, washing powders and many plastic items come from the liquid mineral oil.

Plastics from oil
Plastics are made from an important group of minerals known as liquid hydrocarbons. One of these liquids, known as ethylene, is hardened through pressure and heat to form the solid plastic, polythene.

chopper

spear head

spear head

axe head

arrow head

scraper

Flint tools
Among the first tools made by humans were stone hand axes and blades. One common material for early tool-making was flint. It fractures easily to give a sharp edge and could be flaked to form many different tools.

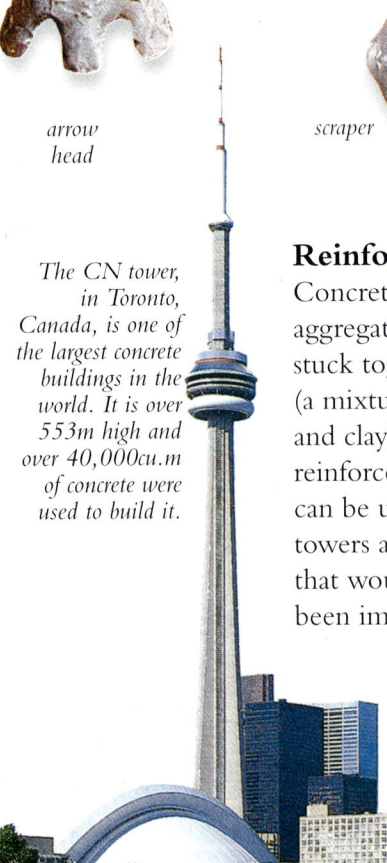

The CN tower, in Toronto, Canada, is one of the largest concrete buildings in the world. It is over 553m high and over 40,000cu.m of concrete were used to build it.

Reinforced concrete
Concrete consists of aggregate (sand and gravel) stuck together by cement (a mixture of limestone and clay). When reinforced with steel, it can be used for high-rise towers and skyscrapers that would otherwise have been impossible to build.

Crushed limestone
In many countries, limestone is used in larger amounts than any other rock. A layer of limestone, crushed into walnut-size pieces, makes a perfect base for tarmac on major roads.

lime

soda ash

sand

kaolin

Coloured pigments also made from minerals are used as decoration on porcelain.

Making glass

High-quality glass is made by melting pure sand, at very high temperatures. Most ordinary glass is made from a mixture of sand with soda ash and lime, because it melts at a lower temperature. Glass was first made by the ancient Egyptians.

Porcelain china

The most highly prized material for making crockery is porcelain because it is strong and waterproof. Other ceramics, such as earthenware, absorb water more easily because they are made from coarser clays. True porcelain is made using very fine china clay, known as kaolin.

Marble slabs

Polished slabs of marble make an eye-catching surface for floors and decorative objects. Marble is a metamorphic limestone, valued because it can be easily cut and polished. Some limestones and granites are popular for the same reasons.

building brick

clay

terracotta pig

Shaping clay

When clay is mixed with water, it becomes malleable (it can be shaped easily). Clay objects are first moulded into shape, then baked in large kilns. Shiny objects are made by coating them with a glaze, which can be made in various colours. Unglazed pottery is called terracotta, which is from the Italian word for baked earth.

FACT BOX

• In the mid-1700s, Coalbrookdale in England became one of the first industrial towns because of its interesting geology. It has an unusual sequence of rocks that includes layers of clay (for pottery and bricks), and coal, iron ore and limestone, which are the essential ingredients for iron-making. Running out of the rocks was natural bitumen, a material that was used to make machine oil. It was also used to waterproof the boats that ▪sported the cast iron.

COAL

peat

lignite or brown coal

black coal

anthracite

The shiny black material that we call coal is a very useful material. It is called a fossil fuel, because it is from the fossils of dead plants. Burning coal in power stations is one of the ways in which electricity is generated. To most geologists, coal is a type of sedimentary rock made of solid minerals. These originally come from plants that died long ago and became rapidly buried by other sediments, commonly sandstones. Much of the world's coal formed from plants that lived and died in the Carboniferous period, which was between 286 and 360 million years ago. At that time, tropical rainforests existed across Europe, Asia and North America. Coal of a different age is found in India and Australia.

Lush rainforest
Damp, swampy rainforest is a similar environment to the one in which coal formed millions of years ago. To produce coal, the remains of plants lay submerged under water, in swamps or shallow lakes, for millions of years.

Types of coal
The hardest, best quality coal is called anthracite. More crumbly black coal and lower grade brown coal (lignite) do not give as much energy when burned. Lignite was formed more recently than black coal and anthracite. Peat is younger still. It forms even today.

Coal seams
Coal is found in layers known as seams. The largest may be several metres thick. Between the seams there are layers of sandstone or mudstone. This machine is extracting coals from a seam at the surface.

Bark fossil
This piece of coal clearly shows the bark patterns of the plant it came from. Plants such as tree-like horsetails, primitive conifers and giant ferns grew in huge forests.

Victorian jet brooch

Since Bronze Age times, jet has been carved into many decorative objects. It was popular in Victorian Britain for mourning jewellery, which was worn in memory of the dead.

Stoking the fire

Coal releases large amounts of energy as heat when it burns. A steam locomotive uses heat from burning coal to produce steam to drive its wheels. The driver keeps the fire well stoked.

Black as jet

Jet is a material that is similar to coal. It is formed from pieces of driftwood that settled in mud at the bottom of the ocean. It is very light and is sometimes polished and carved into intricate shapes.

natural jet

Digging out the coal

In the past, coal was dug from deep underground mines. Today it is mostly taken from large, open pits using gigantic earth-moving equipment. After the coal has been removed the holes left behind are turned into recreation areas. Lakes are created for boating and fishing, and old areas of waste rock are planted with grass and trees. Years ago, this restoration work was not always carried out, so that coal mining areas were unpleasant places in which to live.

At the coal face

A small amount of coal is still mined underground. Tunnels are built so that people and machines can reach an exposed seam of coal, called the coal face. In areas that are difficult to reach, the coal is mined by hand.

Coal-fired power station

Power stations are often built near coal fields. Coal is used to heat water to make steam. Steam-driven turbines run huge generators that produce electricity.

LIQUID MINERALS

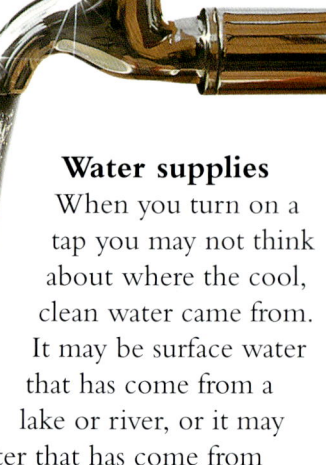

Minerals in rocks also occur in liquid form. Of all the minerals on Earth, one is used by you more than all the others – water. This liquid mineral has greatly influenced the way Earth developed. Without water there would be no oceans, no rivers or lakes, no rain clouds, and no plants or animals. The atmosphere would be quite different. It would probably be made of carbon dioxide, similar to all the other dry, lifeless planets in the Solar System.

Water is constantly being recycled, moving between the atmosphere, the oceans and the rocks of the Earth's crust. As it moves through the crust, it dissolves, grinds, freezes and thaws, changing the rocks it passes through.

In the oceans, living plants and animals have played a part in the making of another important group of minerals. These are liquid hydrocarbons, from which many essential products such as petrol, lubricating oils and natural gas are made.

Water supplies

When you turn on a tap you may not think about where the cool, clean water came from. It may be surface water that has come from a lake or river, or it may be groundwater that has come from deep inside the Earth. Both sources are stopping places in the great cyclic journey made by water. Occasionally water from different points in the water cycle is used. People living on small ocean islands may collect rainwater for their needs. In some countries where rainfall is low, seawater is turned into drinking water by removing the salt.

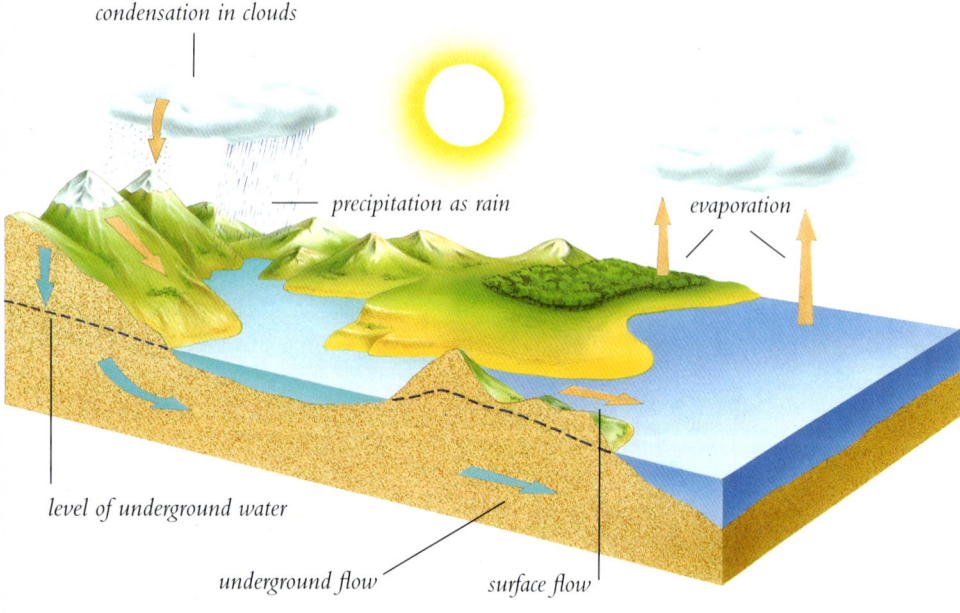

condensation in clouds

precipitation as rain

evaporation

level of underground water

underground flow

surface flow

Water's circular journey

As water moves through its cycle, it changes from water vapour in the atmosphere to drops of liquid rain. Some of the rain falls into rivers and streams which flow into the sea. This water is known as surface water. Some water percolates downwards through cracks and pores in rock and becomes groundwater. Sometimes, groundwater seeps out at the surface in springs, but usually it has to be pumped up through wells or boreholes. Throughout its cycle, water slowly reacts with rock and dissolves elements from it. Two types of water are most commonly found, hard water and soft water. The first contains minerals, the second does not.

Clean water

Surface water, such as river water, is often cloudy because it contains particles of sediment such as clay. Water that has not been cleaned at a water plant also contains microscopic matter that can cause disease. Groundwater contains dissolved elements such as calcium and magnesium, but it is clear and sparkling because the rocks it has been filtered through have cleaned it.

The story of oil begins in warm seas full of living things. As they die, they fall to the sea floor to decay into thick black mud. Unusual things must then happen if oil is to form.

In time, the mud must be buried beneath many layers of sand with clay in between. The sediments must sink deeper and deeper and also become hotter.

After millions of more years, the sediments must come under pressure and fold. Oil from the black mud is then forced into the sandstones and trapped under layers of clay.

From plankton to oil

Liquid hydrocarbons, such as oil, are mixtures of the element carbon and the gas hydrogen. Geologists believe that carbon and hydrogen originally come from small sea creatures known as plankton. When the plankton die, their remains accumulate on the seabed. As they become buried in mud they decompose and make droplets of oil. With time, the mud is buried under very thick layers of sediment. In some situations, the droplets join together to make big underground pools. These get bigger and bigger until they form an oil reservoir.

Extracting oil

Oil reservoirs are often found beneath the sea. Rigs are built over the reservoir to pipe the oil and gas to shore. The rig shown above, in the North Sea, is built on legs that rest on the sea bed. The men who work on the rig are carried to and from shore by helicopter.

The hazards of producing oil

Oil in a natural reservoir is under pressure from whatever is above it. It is usually held in place by a layer of non-porous rock immediately above, which keeps it from seeping away. This is known as the cap rock. When an oil well is drilled through the cap rock into the reservoir, the pressure forces the oil into the well. Strong valves must be fitted to control the flow of oil and turn it off when required. In the well shown on the left, in Kuwait, there was an accident causing a jet of burning oil to rise high into the air. It was eventually extinguished and the well made safe again.

METALS FROM ROCKS

Among the most important minerals are metal ores. These contain minerals rich in metallic elements, such as iron, copper and tin. To extract the metal from its ore, the ore must first be separated from the rock in which it is found. The ore, must then be heated with various other substances in a process known as smelting.

A few metals exist in rocks as a pure element. Gold is the best example of this and is found as veins and nuggets in many types of rock. Gold is dug from the rock in mines, although in some parts of the world it can be found as grains in river sand. The largest grains are known as nuggets.

Metals are useful to people, because they last for a long time. They can also be shaped into many objects or drawn out into a fine wire that will conduct (let through) heat and electricity.

Reclaimed by nature
Metals in the Earth exist joined to other elements. Some quickly rejoin after extraction. Iron quickly rejoins with oxygen and water to form iron oxide, better known as rust.

Metal sculpture
In many cities you will see statues made of metal. Most are made of bronze, a mixture of copper and tin. This famous statue of Eros in Piccadilly Circus, London, is unusual because it is made of aluminium, a silvery metal. At the time the statue was erected in 1893, aluminium was very difficult to make. It is not easily smelted like other metals and is taken from the Earth by using electricity.

Gold Rush
Sometimes rocks surrounding a nugget of gold are eroded by a river. Occasionally, the gold is released into the river and can be recovered through a hand-sieving process called panning. This picture shows the 1849 Californian Gold Rush, when thousands hoped to make a fortune from gold.

Fine gold wires
In this close-up view of a microprocessor, you can see very thin wires of gold. Most metals can be drawn into fine wires – a property called ductility. Gold is the most ductile of all metals. It also resists the corrosive effects of many chemicals, making it an important metal to the electronics industry and in dentistry.

stainless steel cutlery

FACT BOX

• A mixture of two or more metals is called an alloy. The first alloy people made was bronze (a mixture of copper and tin). Bronze is strong and durable and was used to make tools and weapons as long as 6,000 years ago.

• Gold is a rare and precious metal. In the hope of finding gold, some people have mistaken other minerals for it. These minerals are known as fool's gold and include shiny pyrite and brassy-coloured chalcopyrite.

haematite (iron ore)

carbon

Copper

Water pipes and plastic-coated electrical wire are two of the important uses for copper. The metal is taken from its ore first by removing elements that are not copper. Then it is heated in a furnace with a blast of oxygen.

copper pipes

native copper

Iron

In iron ore, atoms of iron are joined to atoms of oxygen. To produce iron metal, iron ore is heated in a furnace with carbon. The carbon takes away the oxygen atoms, leaving the metal behind. Adding more carbon produces steel, which is harder and less likely to snap or corrode. Steel is the most widely used metallic substance.

aluminium foil

bauxite (aluminium ore)

Aluminium

In aluminium ore, atoms of aluminium are tightly joined to atoms of oxygen. Powerful currents of electricity are used to separate the aluminium from the oxygen.

Smelting iron

Iron ore is heated in a blast furnace to produce nearly pure iron. In a process known as smelting, the blast furnace blows hot air through to remove impurities from the ore.

ROCKS AND BUILDINGS

Rocks have been used for building ever since people discovered ways of cutting blocks of stone from the ground. Stonehenge, in south-west England, is evidence that about 4,500 years ago people knew how to cut, shape and move giant slabs of rock into position. Exactly how they did this is a mystery. In South America, the Incas were building large stone buildings long before the voyage of Christopher Columbus in 1492. All over the world, many examples of stone buildings can be seen. Over the last hundred years natural stone gave way to man-made brick and concrete, but today stone is again being used in buildings.

In the past, builders used whatever rock they found close at hand, giving many towns and cities an individual character. For large buildings, the blocks of rock must be free of cracks and natural weakness, and splittable in any direction. It is then called freestone. The best freestones are igneous rocks, especially granite. Of the sedimentary rocks, some sandstones and limestones make good freestone, while marble is the best freestone among the metamorphic rocks. If sandstone splits easily along one direction into thick slabs, it may be used as a flagstone for floors or walls. Slate is a metamorphic rock that can be split into very thin sheets and is ideal for roofing.

The Three Graces
This sculpture is by Antonio Canova, thought to be the greatest sculptor of the 1700s. It is made of marble from Italy and shows the detail that skilled sculptors can achieve using this stone. Polished marble has a special attraction because light is able to penetrate it and is then reflected back to the surface by deeper crystals.

The Taj Mahal
This huge tomb, built in 1400, has an outside surface of white marble that seems to change its appearance in different light and weather. Marble has always been a favourite stone of sculptors because of its gleaming white colour. Its fine grain makes it easy to work with.

Roofing slates
Slate is a perfect covering for roofs. It remains almost unchanged by weather and lasts for many years. Slates are made by splitting blocks along natural cracks known as cleavage planes. Today, tiles and plastic have mostly replaced slate as a roofing material. However, it is still used in thick slabs to provide non-slip flooring in large public areas such as airports. The largest slate quarries in the world are in North Wales.

England's earliest stone building

About 4,500 years ago a hard sandstone was used for the largest stones at Stonehenge in south-west England. Like the Incas, the builders managed to shape and lift these very large stones without using metal tools.

St Paul's Cathedral

After the Great Fire of London destroyed a large area of the city in 1666, the architect Sir Christopher Wren was given the task of rebuilding St Paul's Cathedral. It took him over 35 years. He chose to build the cathedral using a white limestone from Dorset, England, called Portland stone. Over 6 million tonnes were used in the new St Paul's and other buildings in London. The stone was carried in barges along the coast and up the River Thames.

A perfect fit

These stones at the city of Machu Picchu in Peru are of white granite. They were shaped by the Inca people who did not have metal tools. Even now, no-one knows how this was done. The blocks fit together so well a knife cannot be pushed between them. They are even earthquake-proof. The Incas used stone available in the area, such as limestone, rhyolite and andesite.

Walls without cement

In many upland areas where trees do not grow well, stone is used for field boundary walls rather than hedges. They are made with whatever stone blocks can be found close by. Great skill is needed to interlock the different-sized blocks together, without the use of a binding substance, such as cement.

Stone working today

These huge blocks are of white metamorphic marble which is found in the Carrara Mountains of Italy. The blocks are cut from the quarry by drilling lines of closely spaced holes which are then filled with explosive. Carrara marble has a very fine-grained texture and a beautiful lustre when it is polished.

ROCKS IN SPACE

WE have evidence that rocks and minerals exist on other planets. Other planets and rocky material are in orbit around the Sun. They were formed when the Solar System was created, some 4,600 million years ago. Some fragments of the rocky material exists as small particles called meteoroids. Every day, tonnes of this material hit the top of the Earth's atmosphere. Here friction causes it to heat up and vaporize, sometimes causing a spectacular display called a meteor shower, or shooting star. Larger particles do not vaporize completely and a few actually hit the ground. These rocks are called meteorites.

Some meteorites are from the Moon or Mars. They were chunks of rock that were thrown off the planet when rock fragments from Space bombarded the surface, forming craters. The surface of the Moon is littered with craters. When meteorite craters have formed on Earth, they have usually been covered over or destroyed by geological processes, such as the formation of mountains, or erosion by the weather.

*chondrite
(stony) meteorite consisting mainly of the minerals olivine and pyroxene*

*iron
meteorite
consisting of nickle-iron, which is strongly magnetic*

Meteor crater
This hole in the ground in Arizona, USA, is a crater. It was formed by the impact of a meteorite that fell about 25,000 years ago. About six craters of this size exist on Earth. Most craters are covered up, filled with water, or were eroded long ago.

*shergottite
(stony) meteorite consisting of two different basalt rocks*

Meteorites
There are three main types of meteorite – stony, iron and stony-iron. Stony meteorites (the most common) are made of rock. The others contain nickel and iron.

Craters on the Moon
Many millions of years ago the Moon was constantly bombarded by meteorites that hit its surface, making huge craters. The Earth may once have looked like this. It has few craters today because of the action of wind and rain and movements of the crust that makes up the Earth's surface.

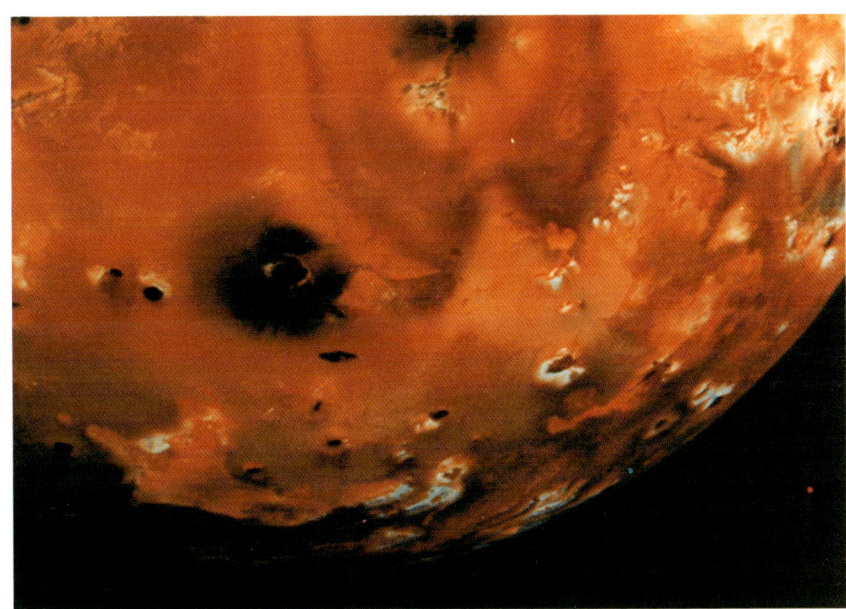

Rocks on Mars
The Pathfinder mission landed on Mars in July 1997. Aboard was a robot probe, called the *Sojourner Rover*, which studied rocks on the surface and sent back photographic images.

Jupiter's moon
This picture, taken in 1979 by the spacecraft *Voyager 1*, shows Io, one of Jupiter's moons. It clearly shows that other planets in the Solar System have volcanoes. However, none of the planets is known to have an outer layer that moves over a liquid centre, as on Earth.

The red planet
Mars is called the red planet because its surface is covered with red iron-oxide dust. It is the most Earth-like of the planets and may hold important clues for Earth's future climate. Four huge volcanoes and an enormous canyon scar its dry surface.

Asteroids
Many meteorites are thought to be broken fragments formed by the collisions between asteroids (small planets). Most asteroids orbit the Sun in a belt between Mars and Jupiter. Inside an asteroid is a central core of metal, which is surrounded by rock.

FACT BOX
• Astronauts have brought about 382kg of Moonrock to Earth. The most common type of rock on the Moon is basalt. It is the same as the basalt that occurs on Earth and formed from solidified lava from volcanoes.

• Only one person is ever known to have been hit by a meteorite. It happened in 1954, in Sylacauga, Alabama. The person was not hurt because the meteorite had already bounced on the ground.

Shooting stars
Meteoroids (small objects from space that hit the Earth's outer atmosphere) travel at high speed. As they pass through our air, they heat up and glow yellow-white, appearing as a bright streak across the sky. This is called a meteor, or shooting star.

STONES FOR DECORATION

Rock paintings
Aboriginal Australians first drew rock paintings like this one thousands of years ago. They used earthy-toned mineral pigments, such as umber (red-brown) and ochre (dark yellow).

raw umber   *brown umber*   *yellow ochre*

Pigments
Mineral colours have been used for thousands of years as pigments. Rocks containing colourful minerals are ground down, mixed with a binder, such as egg yolk, fat or oil, and used as paint.

MINERALS that are highly prized for their beauty are called gemstones. The main use of gemstones is in jewellery or other decorative work, although some are also used in industry. Around 90 minerals are classed as gems. About 20 of these minerals are considered important gems, because of their rarity. These include diamonds, the most valuable of all gemstones. Some minerals provide more than one type of gem. For example, different types of the mineral beryl form emerald, aquamarine, heliodor and morganite. Gems such as ruby and emerald are distinctive because of their deep colour. The different colours of gemstones are caused by metal impurities in the mineral. Other minerals, not necessarily gemstones, are prized for their colour and ground down to pigments. These can be mixed with water to make paint.

The Millennium Star Diamond
This magnificent pear-shaped diamond from South Africa is one of the finest ever discovered. The hundreds of faces have been cut with great skill to bring out the perfection of the stone. It was made to mark the new millennium and is on display in the Millennium Dome in London, England.

Jewellery
This necklace is made of gold set with many precious gemstones. Beautiful minerals have been used for thousands of years in decorative jewellery.

Cameo
The gemstone agate occurs in layers of different colours. This makes it possible to carve in layers, using a decorative technique known as cameo. In a cameo, the top layer is carved to reveal the lower one as a background. This Greek cameo is of Alexander the Great.

Lapis lazuli

A mixture of the minerals lazurite, pyrite and calcite forms lapis lazuli. Its blue colour is caused by the presence of sulphur in the mineral lazurite. The ancient Persians were the first to crush the rock and use the pigment for ultramarine paint.

Specks of pyrite in lapis look like gold.

natural lapis lazuli

Amber beads

Amber is called an organic gem because it is formed from prehistoric tree sap. Most gem-quality amber comes from the shores of the Baltic Sea.

Jade carving

The characteristic green colour of jade comes from atoms of iron metal. Two different minerals are called jade, jadeite and the more common nephrite.

natural nephrite

Blue john is cut and polished to show off the lacy banding.

ruby

diamond

Blue john

The distinctive purple banding in the mineral fluorite is commonly known as blue john. In fact, the bands vary in colour from purple-blue to yellow. Fluorite is a common mineral and is found in limestones.

natural blue john

Blood and fire

A diamond's fiery brilliance, hardness, purity and rarity makes it the most valuable gem. Rubies are rare forms of the mineral corundum. Their blood-red colour comes from the metal chromium.

USING ROCKS AND MINERALS

Many of the materials we use everyday are made from rocks and minerals. Pottery mugs, tin cans and glass windows are just three examples. Many paints, especially those used by artists, are made from colourful minerals. You can make your own paints by crushing rocks in a pestle and mortar.

Gold is one of the few metals that is found in its pure state in nature. It is sometimes found as nuggets in rivers. The nuggets can be separated from mud and gravel by panning, and you can try this in the first project below. In the second project, you can experiment with one of the most widely used mineral materials – concrete.

Make paint
Using a pestle and mortar, crush charcoal, brown clay or chalk. Add oil to the powder to make paint.

You will need: *gloves, trowel, soil, old wok, water, measuring jug, nuts and bolts preferably made of brass, deep tray.*

Gold nuggets
A skilled panner can find single pieces of gold in a whole pan of dirt.

nuggets of "gold"

PANNING FOR GOLD

1 Put on gloves and fill a trowel with soil. Place the soil in an old wok, or shallow pan, along with about a litre of water and some small brass nuts and bolts. Mix it together thoroughly using the trowel.

2 Swirl the wok around, allowing the soil and water to spill over the edge. Add more water if any soil remains and repeat until the water is clear. Panning for real gold washes away the mud, leaving gold behind.

3 When the water is quite clear, examine what is left. You should see that the nuts and bolts have been left behind, as real nuggets of gold would have been, because they are heavier.

PROJECT

MIXING YOUR OWN CONCRETE

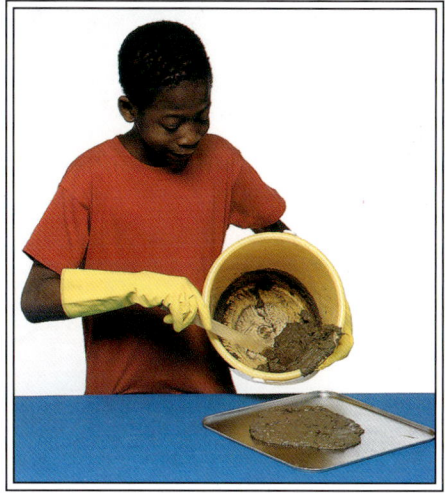

1 Place one cupful of sand in a bucket. Add two cups of cement and a handful of gravel. Don't touch the cement with your bare hands.

2 Add water to the mixture little by little, stirring all the time until the mixture has the consistency of porridge. Mix it well with the stick.

3 Pour the wet concrete on to a tray and spread it out. Leave to solidify for about half an hour. Wash all the other equipment straight away.

You will need: *gloves, measuring cup, sand, bucket, cement, gravel, water, stirring stick, tray, small thin box, foil.*

hand prints

6 How strong is the concrete block once it has set? Test its strength by trying to bend it. Can you rest a heavy weight on the block?

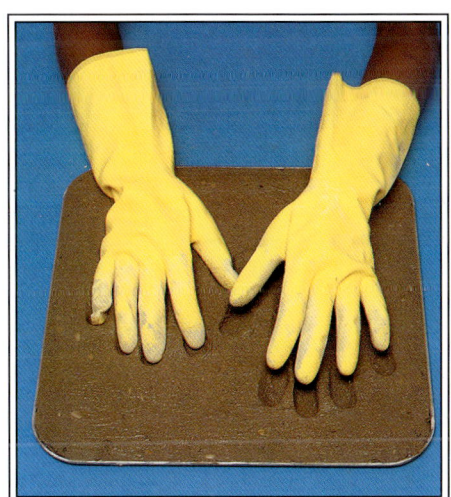

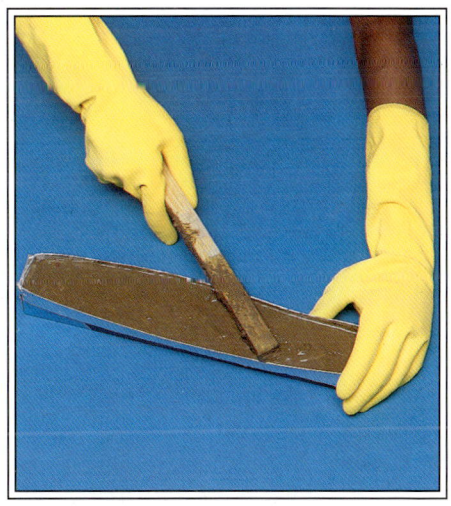

4 At this stage you can shape the concrete any way you like. Make impressions of your hands, or write with a stick. The marks will be permanent once the concrete has set. Do not use your bare hands.

5 You could make a solid concrete block like those used in the construction industry. First, line a small but strong box with kitchen foil. Pour in the concrete and smooth the top with a piece of stick.

VOLCANOES AND EARTHQUAKES

*Volcanoes are among the world's most awesome
natural spectacles. When a volcano erupts,
red-hot molten rock may gush out of the ground.
Fiery cinders may be blasted high into the air,
and superheated steam may burst through rocks.
Volcanoes can be so terrifying –
and so devastating – that people once
thought they were the vengeance
of angry gods. Now, at last, volcanologists
are beginning to discover how they work
and what makes them erupt.*

Author: Robin Kerrod
Consultant: John Farndon

FIRE FROM BELOW

At this moment in various parts of the world volcanoes are erupting. Fountains of red-hot rock are hurtling high into the air and rivers of lava are cascading down the volcanoes' sides. Volcanoes are places where molten (liquid) rock pushes up from below through splits in the Earth's crust. They may be beautiful but they can also be very destructive. Earthquakes are another destructive part of nature. Every year violent earthquakes destroy towns and kill hundreds, sometimes thousands, of people. The constant movements that take place in and beneath the rocky crust that covers the Earth cause volcanoes and earthquakes. The word volcano comes from the name that the people of ancient Rome gave to their god of fire. He was called Vulcan. Volcanology is the term given to the study of volcanoes and the scientists who study them are known as volcanologists.

The greatest
An artist's impression of the massive eruption of the volcano Krakatoa, near Java in south-east Asia, in 1883.

From the depths
When a volcano erupts, magma (red-hot molten rock) forces its way to the Earth's surface. It shoots into the air along with clouds of ash and gas, and runs out over the sides of the volcano. In time layers of ash and lava build up to form a huge cone shape.

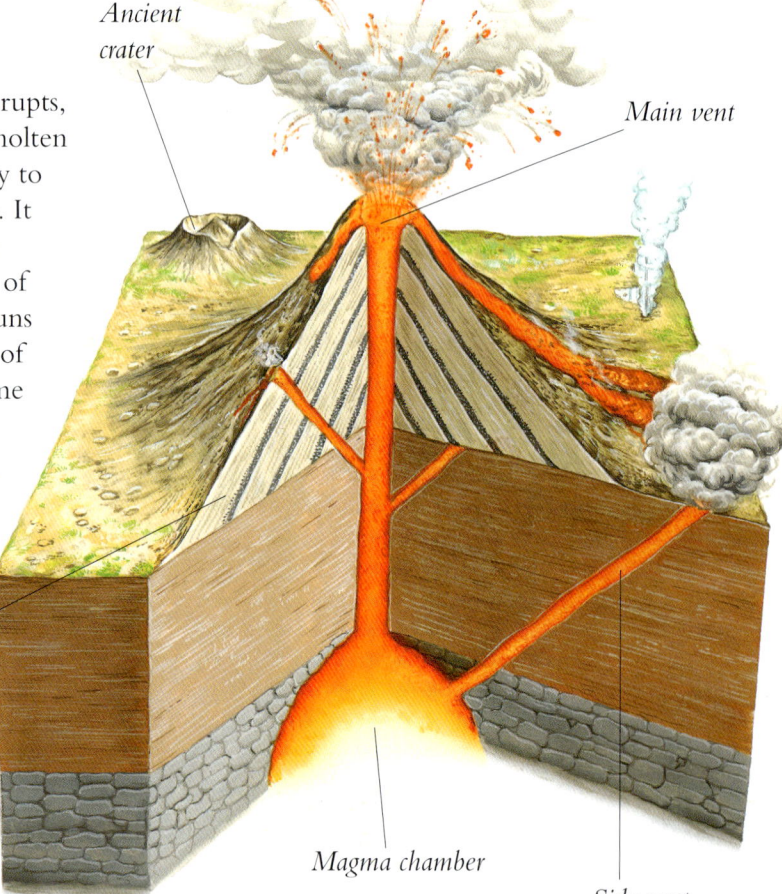

Ancient crater

Main vent

Layers of lava and ash

Magma chamber

Side vent

FACT BOX

• The explosion of Krakatoa in Indonesia, in 1883 caused a massive tidal wave that killed 36,000 people.

• A powerful earthquake killed 5,500 people in the city of Kobe, Japan, in January 1995. Kobe was one of the largest ports in Japan and thousands of homes were destroyed. It was estimated it would take several years to rebuild the city completely.

Piping hot
Hot water often bubbles to the surface in volcanic regions. This creates geothermal (heated in the earth) springs. The hot spring pictured is in Yellowstone National Park in the USA.

Suited up
Heatproof suits and helmets like this make it possible for volcanologists to walk near red-hot lava. This volcanologist is taking samples of lava on the volcano Mauna Loa, on Hawaii.

Out of this world
There are huge volcanoes like this on the planet Venus. Volcanoes have helped shape many bodies in the Solar System, including Mars and the Moon.

Shaking earth
A badly-damaged village in India after a severe earthquake in 1993. Two plates (sections) of the Earth's crust meet in India. The plates push against each other and cause earthquakes.

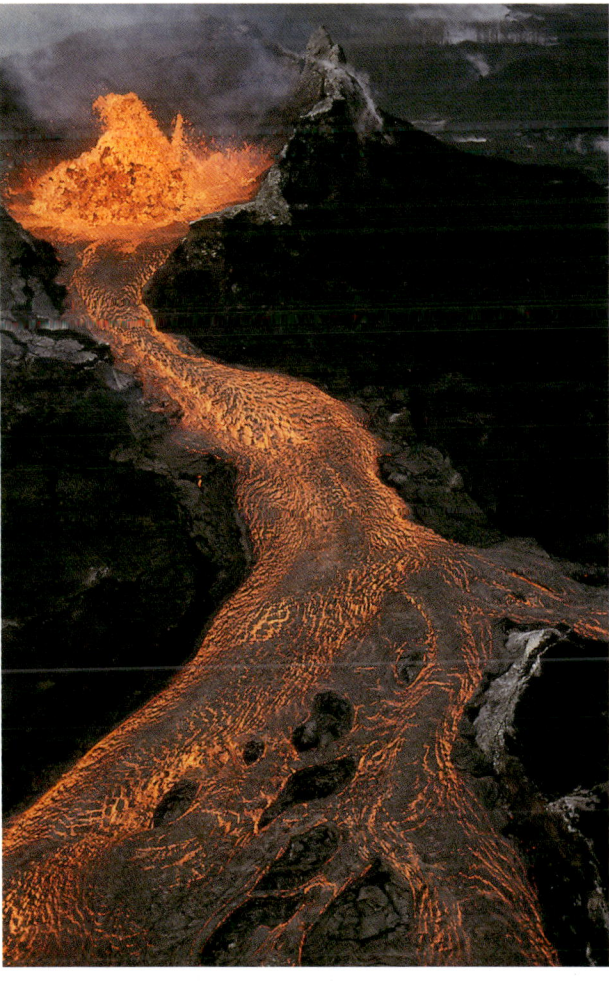

Red-hot river
A river of molten rock races down the sides of the Hawaiian volcano Kilauea in a 1994 eruption. Kilauea is one of the most active volcanoes known on Earth.

THE ACTIVE EARTH

The causes of volcanoes and earthquakes begin many kilometres beneath the surface of the Earth. Our planet is covered with a thin layer of hard rock called the crust. Soil, in which trees and plants grow, has built up on top of the rock. Underneath the hard rock of the crust, however, there is a much hotter layer of the Earth called the mantle. The centre of the Earth, deep inside the mantle, is intensely hot. That heat moves out from the centre and heats everything in the mantle. In the mantle the rocks become semi-liquid and they move and flow like treacle. Because of the intense heat from the centre of the Earth, the rocks move in currents. Very hot liquid rocks (magma) are lighter than cooler rocks and float up towards the top of the mantle. Where there are gaps in the crust, the magma bubbles up through them and shoots out in volcanoes.

Volcanoes have been erupting on Earth for billions of years. During all that time they have pushed out enormous amounts of lava (magma pushed out of a volcano), ash and rocks. These hardened and built up in layers to form part of the landscapes around us. Volcanoes also produced water vapour that eventually condensed (turned to liquid) to form the Earth's seas and oceans.

The newborn Earth
Thousands of millions of years ago the Earth probably looked similar to the picture above. Molten rock was erupting from volcanoes everywhere on the Earth's surface, creating huge lava flows. These hardened into rocks.

Fit for giants
Looking like a spectacular, jumbled-up stairway, this rock formation is on the coast of County Antrim in Northern Ireland. It is known as the Giant's Causeway, because people in the past believed that giants built it. However, it is a natural formation made up of six-sided columns of basalt, one of the commonest volcanic rocks. Basalt often forms columns like these when it cools, and this is called columnar basalt. There are similar structures on Staffa, an island in the Inner Hebrides group off north-western Scotland. Among the many caves along Staffa's coastline is Fingal's Cave, about which the composer Mendelssohn wrote a famous overture.

Inside the Earth

The Earth is made up of a number of different layers. The top layer is the hard crust. It is thinnest under the oceans, where it is only some 5–10km thick. Underneath the crust there is a thick layer of semi-liquid rock known as the mantle. Beneath the mantle is a layer of liquid metal, mainly iron and nickel, that makes up the Earth's outer core. The inner core at the centre is solid, made up of iron and other metals.

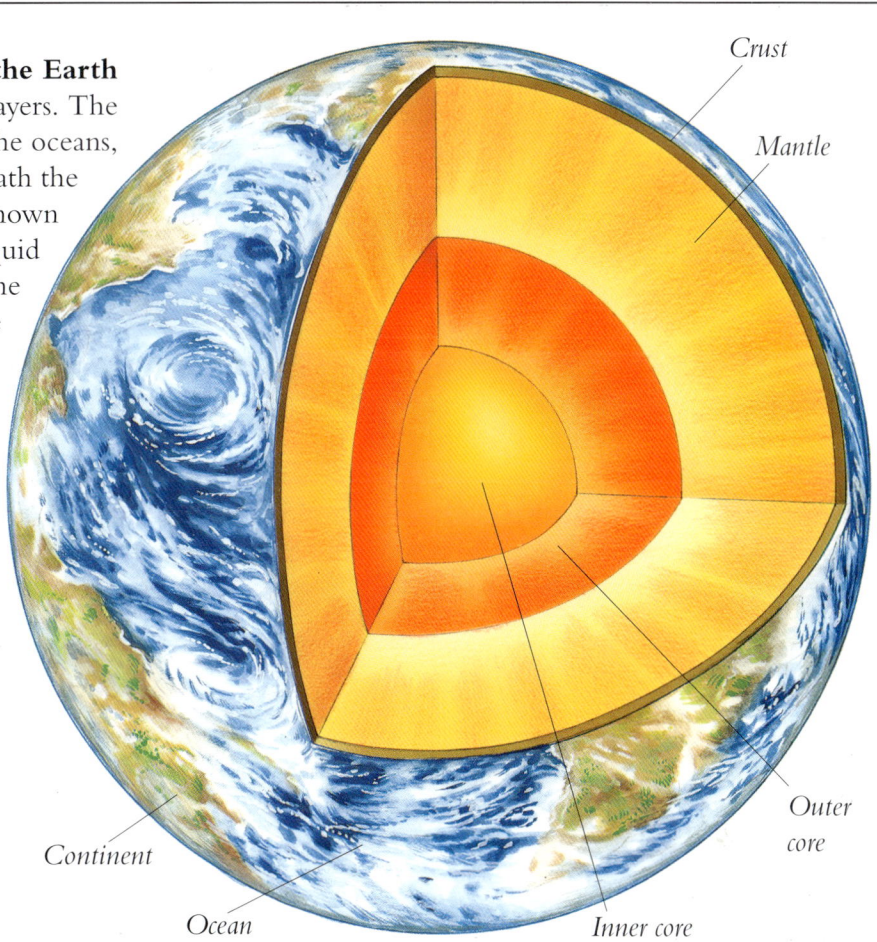

Crust

Mantle

Continent

Ocean

Outer core

Inner core

Fire and ice

This bleak landscape in Iceland was created by the country's many volcanoes. Iceland is one of the most volcanically active places in the world and hardened lava covers most of the country.

Iron from space

Iron is also found in meteorites that fall to our planet from space. This 60-tonne iron meteorite is the world's largest. It was found in 1920 at Hoba, Namibia, in south-west Africa. Scientists estimate that it fell to Earth about 80,000 years ago.

A rocky layer cake

There are other rocks on Earth besides those made by volcanoes. Sedimentary rocks were formed out of sediment, or material produced when older surface rocks were worn away by wind and rain. These kinds of rocks build up in layers. This picture of the Grand Canyon in the USA shows an enormous area of sedimentary rocks. You can see how they are built up in layers of different colours.

ERUPTION

People usually think of volcanoes as mountains of fire that shoot fountains of red-hot rock high into the air and pour out rivers of lava. But much more comes out of volcanoes besides molten rock. Water that has been heated and turned into a gas in the volcano comes out as water vapour and steam. Once it is outside the volcano the vapour cools down and condenses (turns back into water). The hot rock inside volcanoes produces many other kinds of gas, such as carbon dioxide. Some of these gases go into the air outside the volcano and some are mixed with the lava that flows from it. The second project shows you how to make a volcano that gives out lava mixed with carbon dioxide. As you will see, the red floury lava from your volcano comes out frothing, full of bubbles of this gas. In a real volcano, it is the gas that is mixed with the lava that makes the volcano suddenly explode. The gas bubbles and swells inside the volcano and pushes out the mixture of lava and gas violently.

Hawaiian fire
The gigantic Hawaiian volcano Mauna Loa erupted in 1984, sending rivers of red-hot lava cascading down its slopes. The lava came dangerously close to the coastal town of Hilo. If the lava had reached Hilo, the town would have been set on fire.

WATER VAPOUR

You will need: heat-proof jug, saucepan, oven glove, plate.

3 After a few minutes, turn off the cooker and take the plate away using the oven gloves. You will see that the plate is covered with drops of water. This water is water vapour (steam) that has cooled and turned back to liquid.

1 Fill up the jug with water from the hot tap. Pour the water into the saucepan. Switch on one of the hotplates or light a gas ring on a cooker and place the saucepan on it.

2 Heat the water in the saucepan until it is boiling hard and steam is coming from it. Pick up the plate with the oven gloves and hold it upside-down above the saucepan.

P R O J E C T

1 Make sure the jug is dry, or the mixture will stick to the sides. Empty the baking soda into the jug and add the flour. Thoroughly mix the two using the stirrer.

ERUPTION

You will need: jug, baking soda, flour, stirring rod, funnel, plastic bottle, sand, seed tray (without holes), large plastic bin lid, vinegar, red food colouring.

5 Pour the vinegar into the jug. Then add enough food colouring to make the vinegar a rich red colour. White wine vinegar will make a richer colour.

2 Place the funnel in the neck of the plastic bottle. Again, make sure that the funnel is perfectly dry first. Now pour in the mixture of soda and flour from the jug.

6 Place the funnel in the mouth of the plastic bottle and quickly pour into it the red-coloured vinegar in the jug. Now remove the funnel from the bottle.

3 Empty sand into the tray until it is half-full. Fill the jug with water and pour it into the tray to make the sand sticky but not too wet. Mix together with the stirring rod.

4 Stand the bottle containing the flour and soda mixture in the centre of the plastic lid. Then start packing the wet sand around it. Make the sand into a cone shape.

7 The sandy volcano you have made will begin to erupt. The vinegar and soda mix to give off carbon dioxide. This makes the flour turn frothy and forces it out of the bottle as red lava.

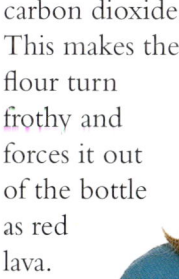

SPREADING SEAS

The Earth's crust is not all in one piece. but is made up of many sections, called plates. They are solid and float on currents circulating in the deep layer of semi-liquid rock beneath them in the mantle. All the plates of the crust move in different directions. Some plates are moving apart and others are moving towards one another. The plates that move apart are usually under the oceans. Magma pushes up from below the sea-floor and squeezes through gaps between the edges of the ocean plates. The magma squeezing up pushes the plates in opposite directions. As the plates move apart they make the ocean floor wider and push continents apart. This is known as sea-floor spreading. Sea-floor spreading builds up plates because when the magma cools it adds new rock at the edges of the plates. This kind of boundary between two plates is called a constructive boundary. The magma pushes up the seabed to form a long mountain range called a ridge or rise. Ridges are very noticeable features on the floors of both the Atlantic and the Pacific Oceans.

Underwater explorer
The deep-diving research submersible (midget submarine) *Alvin*. The submersible can carry a pilot and two scientific observers to a depth of 4,000m. *Alvin* can dive so deeply that scientists were able to study the Mid-Atlantic Ridge. The submersible also helped scientists to discover mineral deposits on the ocean floor.

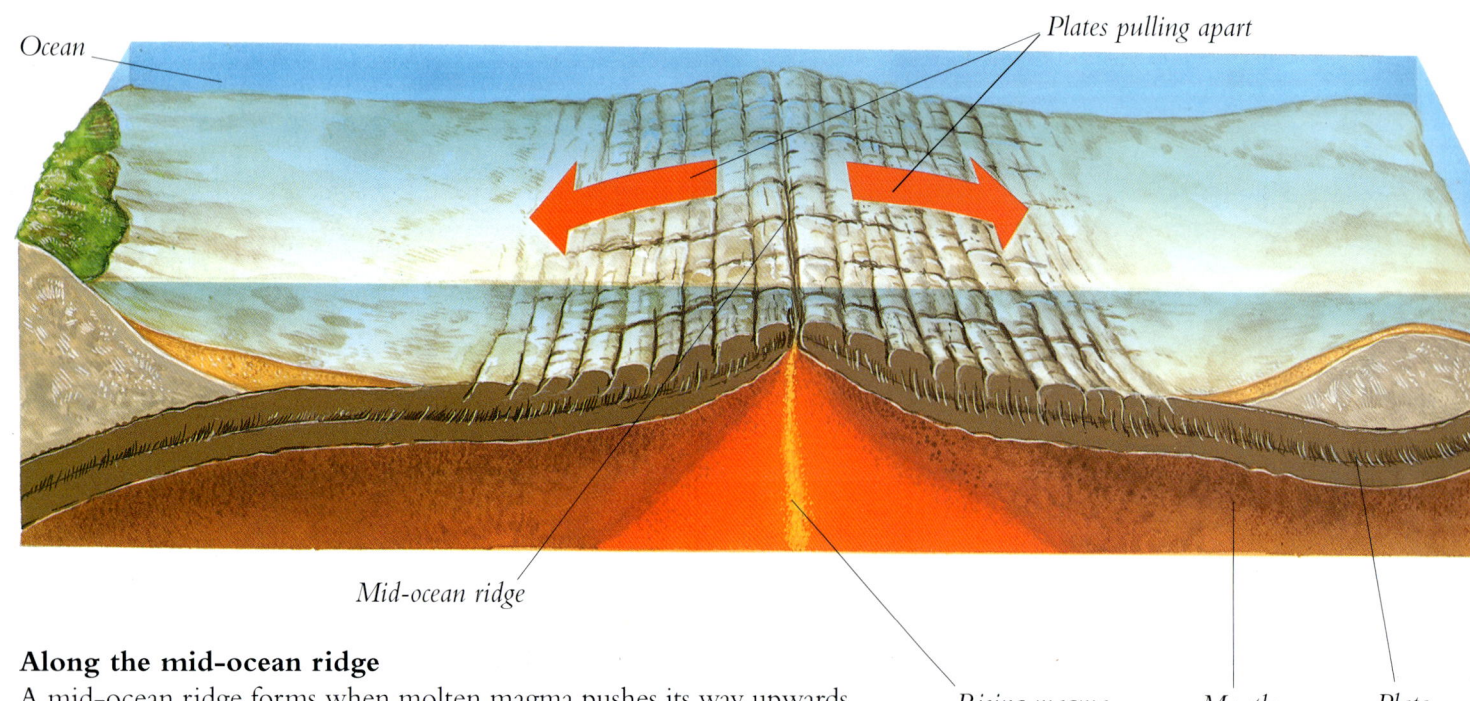

Ocean

Plates pulling apart

Mid-ocean ridge

Rising magma   *Mantle*   *Plate*

Along the mid-ocean ridge
A mid-ocean ridge forms when molten magma pushes its way upwards from the mantle, the semi-molten layer under the crust. The magma bubbles up through cracks in the crust as they are pulled apart. When the magma meets the sea water it hardens to form ridges.

Strange life

Hot water full of minerals streams out of the vents (openings) along the mid-ocean ridges. Strange creatures live around them. They include the giant tube worms pictured here around vents in the Galapagos Islands in the eastern Pacific Ocean. Other creatures that thrive on the ridges include species of blind crabs and shrimps.

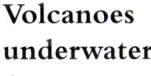

Volcanoes underwater

A computer-coloured picture of three underwater volcanoes in the South Pacific Ocean. They were found close to the East Pacific Rise, the main mid-ocean ridge in the Pacific Ocean.

New island

The island of Surtsey, off Iceland, did not exist before November 1963. In that month the top of an erupting volcano broke through the sea's surface close to Iceland. The volcano continued erupting for more than three years. There were times when the ash and steam rising from the new volcano reached more than 5km into the sky.

Black smoker

Water as hot as 350°C, hotter than a domestic oven, pours out of vents along the ocean ridges. The water often contains particles of black sulphur minerals that make it look like smoke. This is why these vents are called black smokers.

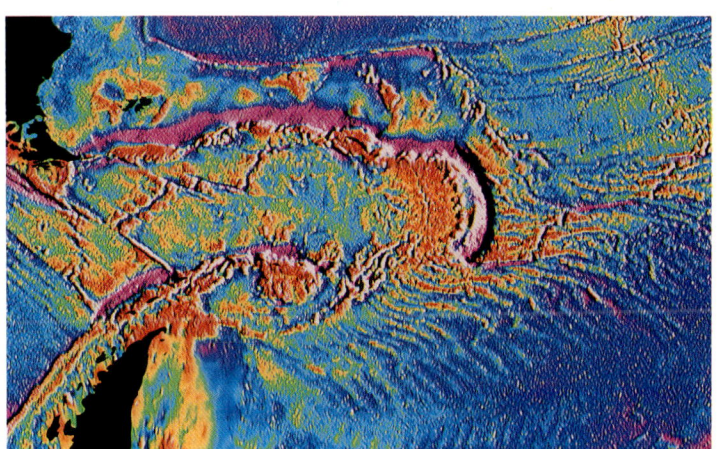

Plate and sandwich

A satellite view showing a small ocean plate on the floor of the South Atlantic Ocean between the tip of South America (top left) and Antarctica (bottom left). The curved shape in the centre is the South Sandwich Trench.

PROJECT

MOVING MAGMA

PLASTIC FLOW

You will need: lump of modelling clay, wooden board.

1 Make sure that the table is protected by a sheet. Knead the lump of clay in your hands until it is quite flexible. Now shape it into a ball. Place it on the table.

2 Place the wooden board on top of the ball of clay and press down. The clay flattens and squeezes out at the sides. It is just like semi-liquid rock flowing under pressure.

The temperature of the rocks in the Earth's mantle can be as high as 1,500°C. At this temperature the rocks would normally melt. They are under such pressure from the rocks above them that they cannot melt completely. They are, however, able to flow slowly. This is like a solid piece of modelling clay that flows slightly when you put enough pressure on it. This kind of flow is called plastic flow. In places, the rocks in the upper part of the mantle do melt completely. This melted rock, called magma, collects in huge pockets called magma chambers. The magma rises because it is hotter and lighter than the semi-liquid rocks. Volcanoes form above magma chambers when the hot magma can rise to the surface. The second project demonstrates this principle using hot and cold water. The hot water rises through the cold because it is lighter.

Rock currents
Underneath the Earth's hard crust, the rock is semi-liquid and can move slowly. It moves in currents. Hot rock moves upwards and cooler rock sinks back down.

3 Roll the clay into a ball again and press it with the board. But this time push the board forwards at the same time. The clay will again flow and allow the board to move forwards. The board is moving in the same way as the plates in the Earth's crust move.

PROJECT

1 Pour some of the food colouring into the small jar. You may need to add more later to give your solution a deep colour. This will make the last stage easier to see.

2 Fill the small jug with water from the hot tap. Pour it into the small jar. Fill it right to the brim, but not to overflowing. Wipe off any that spills down the sides.

3 Cut a circular patch from the plastic food wrapping a few centimetres bigger than the top of the small jar. Place it over the top and secure it with the elastic band.

BLACK SMOKERS

You will need: dark food colouring, small jar (such as baby food jar), small jug, transparent plastic food wrapping, strong elastic band, sharpened pencil, large jar, oven gloves, large jug.

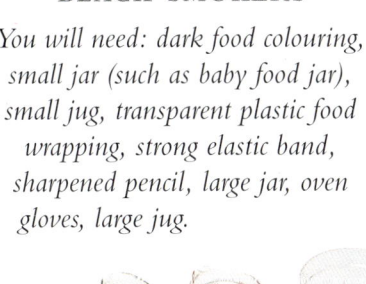

4 With the sharp end of the pencil, carefully make two small holes in the plastic covering the top of the jar. If any coloured water splashes out, wipe it off.

6 Watch what happens. The coloured hot water begins rising from the holes. This happens because the hot water is lighter, or less dense, than the cold water around it.

5 Now place the small jar inside the larger one. Use oven gloves because it is hot. Fill the large jug with cold water and pour it into the large jar, not into the small one.

WHEN PLATES MEET

The plates on the sea floor that spread out from the mid-ocean ridges meet edge-to-edge with the plates carrying the continents. The edges of the plates then push against each other. Because continental rock is lighter than ocean-floor rock, the edge of the continental plate rides up over the edge of the ocean plate. The ocean plate is then forced back into the mantle inside the Earth. As the ocean plate goes down into the mantle, it melts and is gradually destroyed. This kind of boundary between colliding plates is called a destructive boundary because the edge of the ocean plate is destroyed. Where the ocean plate starts to descend, a deep trench forms in the sea-bed. The continental plate is also affected by the ocean plate pushing against it. It wrinkles up and ranges of fold mountains are formed. The great mountain chains of North and South America – the Rockies and the Andes – were formed in this way. Earthquakes also occur at destructive boundaries. So do volcanoes, as parts of the destroyed ocean plate force their way through the weakened continental crust.

Around the Pacific

Most of the Pacific Ocean sits on one huge plate moving north-west. This rubs against other plates and creates a huge arc of volcanoes (shown in red) along the plate edges.

FACT BOX

• The Andes Mountains of South America were formed by the collision between the South American plate and the Nazca plate. With a length of nearly 9,000km, they form the longest mountain range in the world.

• The deepest part of the world's oceans is Challenger Deep, which lies in the Marianas Trench in the North Pacific Ocean. The depth there is almost 11km.

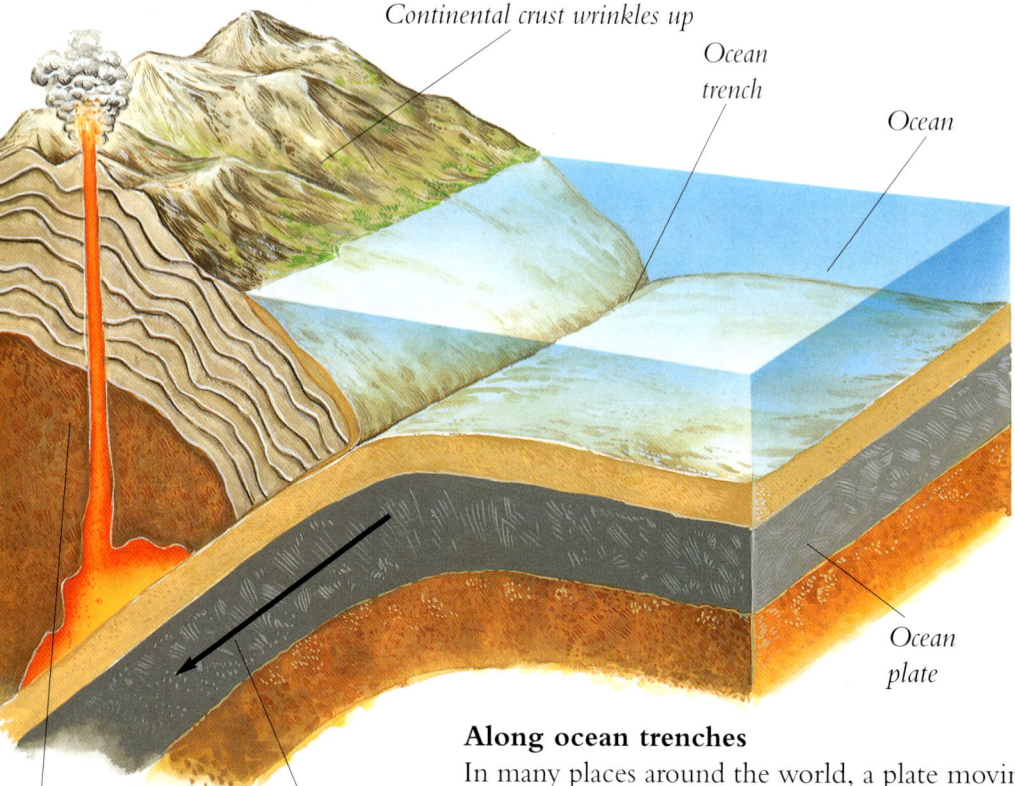

Continental crust wrinkles up

Ocean trench

Ocean

Ocean plate

Continental plate

Ocean plate descends

Along ocean trenches

In many places around the world, a plate moving away from an ocean ridge meets a plate carrying a continent. When this happens, the ocean plate (which is made up of heavier material) is forced down underneath the continental plate. This causes a deep trench to form where the plates meet.

Youngest and tallest

The Himalayas in southern Asia form the highest mountain range in the world. They include Mount Everest which, at 8,848m, is the Earth's highest single peak. The range began rising only about 50 million years ago. The plate carrying India collided with the Asia plate at that time. The Himalayas is one of the youngest mountain ranges on Earth. In the long history of the Earth, 50 million years is not a particularly long time.

The ocean trenches

This satellite photograph of the Earth shows Australia at lower centre with the huge land mass of Asia at the top of the globe. The North and South Pacific Oceans are to the right, while the Indian Ocean is on the left. Variations in the height of the sea surface are clearly visible. The surface dips in places where there are deep trenches on the ocean bed many kilometres below. The deep trenches to the right of Australia in this view are the Kermadek and Tonga Trenches. The Marianas Trench is upper centre.

Volcanoes on the edge

The volcano on White Island, off the North Island of New Zealand lies close to an ocean trench, where the Pacific plate is descending. As the plate descends, it heats up and changes back to magma. This forces its way to the surface as a volcano. White Island volcano is one of hundreds that ring the Pacific Ocean. Together they form a ring of volcanoes which is called the Ring of Fire.

Above the clouds

The tops of volcanoes rise above the clouds on the Indonesian island of Java. Most of the country's islands lie near the edge of a descending plate and have active volcanoes.

HOT SPOTS

Most of the world's volcanoes lie at the edges of plates. A few volcanoes, however, such as those in Hawaii, are a long way from the plate edges. They lie over hot spots beneath the Earth's crust. A hot spot is an area on a plate where hot rock from the mantle bubbles up underneath. While the plate above moves, the hot spot stays in the same place in the mantle. The hot spot keeps burning through the plate to make a volcano in a new place. A string of dead volcanoes is left behind as the plate moves over the hot spot. Some form islands above the ocean surface. Others, called sea mounts, remain submerged. The best known active volcanoes far from plate boundaries are Kilauea and Mauna Loa on the main island of Hawaii. The Hawaiian archipelago lies in the middle of the Pacific plate, thousands of kilometres from plate boundaries. Its volcanoes erupt because it lies directly above a hot spot. The other Hawaiian islands formed over the same hot spot but were carried away by plate movement. In time, the main island will be carried away also. Volcanoes erupting from the hot spot will create a new island to take its place. The island of Réunion in the Indian Ocean is another example of a hot-spot location.

Powerful Pele
This is the name of the fire goddess of Hawaii. According to legend, Pele lives in a crater at the summit of the volcano Kilauea. When she wishes she melts the rocks and pours out flows of lava that destroy everything in their path. When Pele stamps her feet, the Earth trembles.

FACT BOX

• The Hawaiian hot-spot volcano of Mauna Kea is 9,000m high from the ocean floor. That is nearly 1,000m taller than Mount Everest. Half of Mauna Kea is below sea level.

• When sea mount volcanoes die, they cool and shrink. It is possible that the legendary lost city of Atlantis could have been built on top of a flat sea mount. Then the sea mount shrank and Atlantis sank beneath the waves.

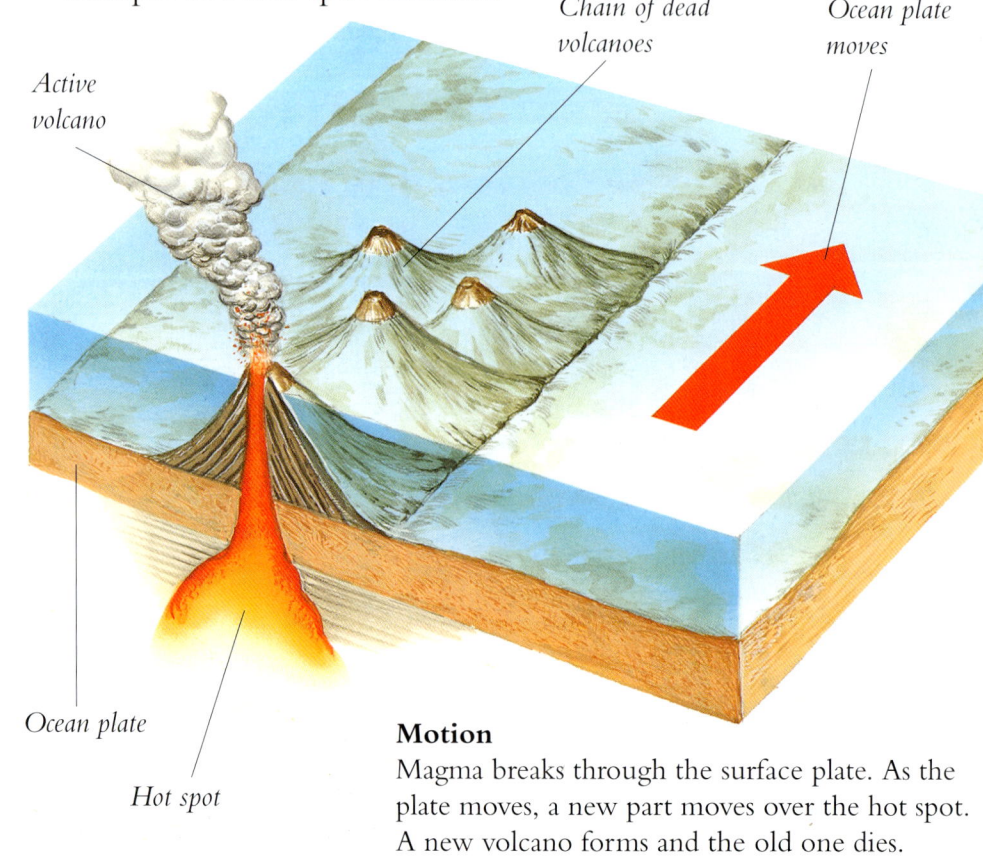

Chain of dead volcanoes

Ocean plate moves

Active volcano

Ocean plate

Hot spot

Motion
Magma breaks through the surface plate. As the plate moves, a new part moves over the hot spot. A new volcano forms and the old one dies.

Islands in line
Astronauts took this picture of the Hawaiian island chain in the North Pacific from the space shuttle *Discovery* in 1998. This island group is formed over a hot spot on the Pacific plate. The largest island is Hawaii which appears at the top of this picture.

Lanzarote's lunar landscape
Huge volcanic eruptions took place on Lanzarote, another Canary Island, in the 1800s and 1900s. They covered most of the island with lava and ash. The landscape is similar to the landscape on the Moon. Very few plants can grow in a landscape of this kind. Lanzarote has more than 300 volcanic craters. Many are to be seen in the area of the most recent lava flows, in the spectacular "Mountains of Fire".

Canary hot spot
Snow-capped Mount Teide, the highest peak on Tenerife, in the Canary Islands. It rises to 3,718m and was formed 10 million years ago by volcanic activity over the Canary Islands' hot spot. Teide last erupted in 1909.

Pacific atoll
There are many ring-shaped coral islands, or atolls, in the Pacific Ocean. These began as coral grew around the mouth of a volcano that rose above the ocean's surface. The the volcano sank, but the coral went on growing.

Pele is angry!
The volcano Kilauea, on Hawaii, is shown erupting here. At such times Hawaiians say that their fire goddess Pele is angry. She is supposed to live in Kilauea's crater. Kilauea, on the main island of Hawaii, is located over a hot spot on the Pacific plate. It formed only about 700,000 years ago.

CONES AND SHIELDS

Around the world there are more than 1,000 active volcanoes. They are all very different. Some erupt fairly quietly and send out rivers of molten lava that can travel for many kilometres. Others erupt with explosive violence, blowing out huge clouds of ash. The kind of magma inside a volcano makes the difference between it being quiet or explosive. Quiet volcanoes, such as those that form on the ocean ridges and over hot spots have magma with very little gas in it. The Hawaiian volcanoes formed over a hot spot. Their lava flows far, and they grow very broad. They are called shield volcanoes. Explosive volcanoes have magma inside them that is full of gas. Gas pressure can build up inside a volcano until it explodes. This is the kind of volcano found in the Ring of Fire around the Pacific plate. Because of their shape these volcanoes are called cone volcanoes. The blast and ash clouds these volcanoes give off can and do kill hundreds of people. The ash clouds can even cause changes in the weather. Large clouds of dust in the Earth's atmosphere from volcanoes block out the Sun's heat, making the weather on Earth colder.

Sticky rock

A volcano erupts with explosive force on Bali. It is one of a string of islands that make up Indonesia. There are more than 130 active volcanoes on the islands. They all pour out the sticky type of lava Hawaiians call aa.

Red river

A river of molten lava flows down the slopes of the volcano Kilauea on the main island of Hawaii. Like the other volcanoes on the island, Kilauea is a shield volcano. It pours out very runny lava that flows for long distances, usually at speeds up to about 100m an hour. The fastest lava flows are called by their Hawaiian name of pahoehoe.

Long mountain

The volcano Mauna Loa on the main island of Hawaii is of 4,170m high. It is a shield-type volcano, meaning it is broad, with gently sloping sides. The main dome measures 120km across and its lava flows stretch for more than 5,000 sq km. In the Hawaiian language, the name means Long Mountain. This is a good name for it because it is very long and is the biggest mountain mass in the world.

Building layers

Explosive volcanoes blast rock and ash into the air. These eventually fall to the ground and lie there. Geologists call the rock and ash on the ground tephra. Here on Mount Teide, in Tenerife, layers of tephra have built up on top of each other after repeated eruptions.

Sacred mountain

Snow-covered Mount Fuji on the island of Honshu, Japan. Also called Fujiyama, it is one of the most beautiful volcanoes in the world and is considered sacred by the Japanese. It has an almost perfect cone shape. Five lakes ring the base of the volcano.

At the top

The caldera (crater) at the summit of the volcano Kilauea, on the island of Hawaii. There are vents (holes) in the caldera from which lava flows. The most active vent in the caldera is named Halemaumau. This is the legendary home of the fire goddess Pele.

Submarine (undersea) volcanoes may grow in size until they rise above the surface of the sea. Scientists believe that this is how atolls are formed.

Hawaiian volcanoes have runny lava and gentle slopes

Strombolian volcanoes spit out lava bombs in small explosions.

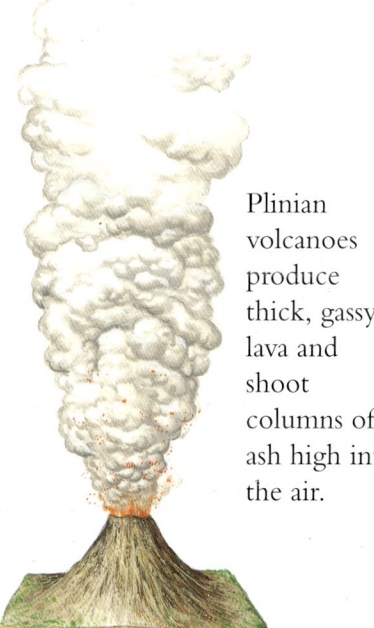

Plinian volcanoes produce thick, gassy lava and shoot columns of ash high into the air.

Fissure volcanoes are giant cracks in the ground from which lava flows.

Vulcanian volcanoes produce thick, sticky lava and erupt with violent explosions.

Pelean volcanoes produce clouds of very hot ash and gases. These clouds are dense and roar or gush quickly downhill.

Volcano types

Although all volcanoes behave in different ways, we can group them into a number of different kinds. In fissure volcanoes, magma forces its way up through long cracks in the Earth's crust. Then it flows out on either side and cools to form broad plateaus. Other volcanoes grow in various shapes caused by how runny or thick their lava is. Some of these volcanoes are famous for their violent eruptions of thick clouds of ash and gas.

PROJECT

FLOWING LAVA

LAVA VISCOSITY

You will need: two paper plates, jar of liquid honey, pen, stopwatch, jug of ordinary washing-up liquid.

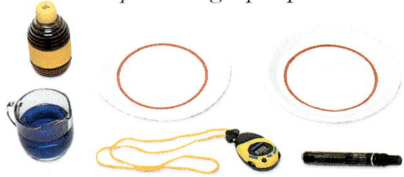

1 Mark a large circle on the plates by drawing around the edge of a saucer. Pour a tablespoon of honey from the jar into the middle of the circle. Start the stopwatch.

In some parts of the world, there are ancient lava flows that are hundreds of kilometres long. Long flows like these have come from fissures (cracks) in the crust, which have poured out runny lava. Runny lava is much thinner than the lava produced by explosive volcanoes, which is sometimes called pasty lava. The correct name for the thickness of a liquid is viscosity. Thin liquids have a low viscosity, thick liquids a high viscosity. The first project below investigates the different viscosities of two liquids and how differently they flow. The second project looks at at the effect on substances of temperature. Heating solids to a sufficiently high temperature makes them first turn soft, then melt and then flow. Rock is no exception to this rule. If you make rock hot enough it softens, becomes liquid, and then flows. Deep inside a volcano, hot rock becomes liquid and flows up and out onto the surface as lava. When the lava comes out, its temperature can be as high as 1,200°C. This is the temperature of most of the runny lavas of the Hawaiian shield volcanoes. There are two kinds of lava flows from these volcanoes. One is called pahoehoe and the other aa by the Hawaiians. Volcanologists use these names for similar flows the world over. Pahoehoe and aa flows have different kinds of surfaces. Pahoehoe has quite a smooth skin and wrinkles up like coils of rope. Aa flows have a very much rougher surface that is full of rubble.

2 After 30 seconds, mark with the pen how far the honey has run. After another 30 seconds mark again. Stop the watch when the honey has reached the circle.

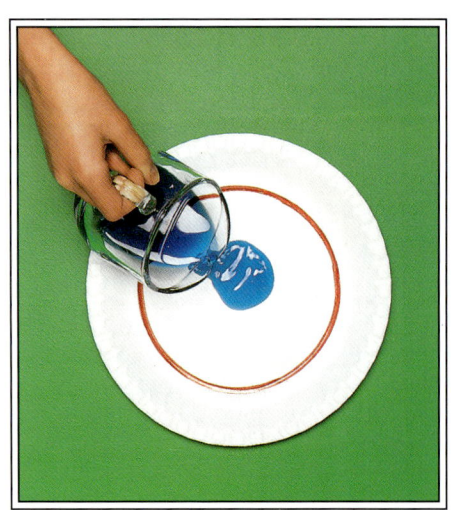

3 Part-fill the jug with washing-up liquid and pour some into the centre of another plate. Use the same amount as the honey you poured. Start the stopwatch.

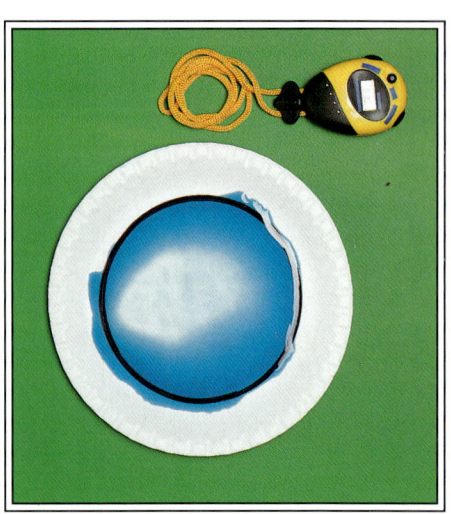

4 After 30 seconds, note how far the liquid has run. You will probably find that it has reached the circle. It flows faster because it has a much lower viscosity than honey.

P R O J E C T

MAGMA TEMPERATURE

You will need: block of hard cooking margarine, jam jar, jug, large mixing bowl, stopwatch.

1 Scoop out some margarine and drop it onto the bottom of the jar. For the best results, use hard cooking margarine, not a soft margarine spread.

2 Pick up the jar and tilt it slightly. See what happens to the margarine. The answer is, not a lot. It sticks to the bottom of the jar and does not slide down.

6 Continue checking the jar for another three or four minutes. After even a minute, the margarine will start to slide along the bottom as it warms and starts to melt. After several minutes, it is quite fluid.

3 Fill the jug with hot water and pour some into the bowl. Shake it around to heat the bowl, then pour it away. Now pour the rest of the hot water into the bowl.

4 Pick up the jar and tilt it again. The margarine still will not move. Now place the jar on the bottom of the bowl. Keep your fingers clear of the hot water.

5 Start the stopwatch and after one minute, take out the jar. Tilt it, and see if the margarine moves. Return it to the bowl and after another minute, look at it again.

VICIOUS VOLCANOES

Erupting volcanoes can be among the most impressive sights in nature. They are almost always destructive, however, and can be deadly. Quiet volcanoes are the least dangerous to life, but their lava flows will destroy anything in their path. Explosive volcanoes are the most destructive. Their lava does not flow far because it is so thick. However, the clouds of ash, shattered rock and gas they blast out can be deadly. It was an exceptionally heavy ash fall that killed people by the thousand in ancient Pompeii. When the volcanoes explode first, they often give off a glowing cloud of white-hot ash, gas, and rocky debris. This is called a *nuée ardente* (glowing avalanche) and can travel at speeds of up to 100 km/h. Such a cloud killed tens of thousands in St Pierre in the Caribbean in 1902. The gases that all volcanoes give out can also be deadly. They include sulphur dioxide and hydrogen sulphide. Both gases are highly poisonous. Vast amounts of carbon dioxide are also given off. This is not poisonous in itself, but it can kill by suffocation (inability to breathe). The carbon dioxide blocks out oxygen. When there is no oxygen, people cannot breathe. In 1986, more than 1,500 people and many animals died in this way at Lake Nyos in Cameroon.

Menace on Montserrat
Ash clouds billow high into the sky from this volcano in the Soufrière Hills on the Caribbean island of Montserrat early in 1997. Many people had to flee from the island.

Long ago in Herculaneum
The excavated remains of one of the houses in the Roman town of Herculaneum, near Naples in Italy. It was destroyed at the same time as nearby Pompeii in August AD79. It was blasted by hot gas and buried by repeated avalanches of hot ash and rock from Vesuvius.

Plaster casts of bodies

The Garden of Fugitives
In this part of the excavated city of Pompeii plaster casts of victims of the AD79 eruption of Vesuvius are displayed. Their life-like casts show how they huddled together in fear.

The death of Pompeii

On 24 August AD79, Mount Vesuvius, near Naples in Italy, erupted with explosive violence. A huge, choking cloud of gas, hot ash, and cinders blew down and covered the Roman town of Pompeii. At least 2,000 people are thought to have been killed either in their homes, or trying to flee from the deadly cloud. In a short time most of the city was buried. Over the past century more than half of the city has been uncovered. The ash and cinders have been dug away from many different buried buildings.

Silent killer

Carbon dioxide killed these cattle in fields near Lake Nyos, in Cameroon. The gas was released during a volcanic explosion under the lake in August 1986.

Lava rain

Volcanic bombs on the slopes of Mount Teide, on Tenerife, in the Canary Islands. They were thrown out during an eruption of the volcano as lumps of partly molten lava.

Gas sampling

A volcanologist takes a sample of gases from a volcanic vent. He wears a gas mask to avoid being suffocated.

Indian burial

The cinder field around Sunset Crater in Arizona, USA. The volcano that created the crater erupted in about AD1064. Thick lava flows, fumaroles (gas vents) and ice caves have been found in the surrounding area. It has been a US national park since 1930.

FACT BOX

• In April 1815 on the island of Sumbawa, in Indonesia, the volcano Tambora exploded. An estimated 90,000 people died directly from the eruption or from famine caused by ruined crops.

• In May 1902, a glowing cloud of gas from the Mount Pelée volcano on the Caribbean island of Martinique destroyed the city of St Pierre and killed its 30,000 inhabitants.

PROJECT

DANGEROUS GASES

The two projects here look at two effects the gases given out by volcanoes can have. In the first project you will see how the build up of gas pressure can blow up a balloon. If you have put enough gas-making mixture in the bottle, the balloon may explode. Be careful. When the gas pressure builds up inside a volcano, an enormous explosion takes place, often releasing a deadly hot gas cloud like the one that killed thousands of people in Pompeii. The second project shows the effect of carbon dioxide, a gas often given out by volcanoes. The project uses the gas to prevent oxygen reaching a candle. The candle cannot burn without oxygen. This explains how carbon dioxide kills people by suffocation: it stops oxygen getting into their lungs. The project also shows that carbon dioxide is heavier than air. Being heavy makes it dangerous because clouds of the gas can push away the air from around people and animals.

GAS PRESSURE

You will need: funnel, drinks bottle, baking soda, vinegar, jug, balloon.

Plaster casts

Gas killed many of those who died at Pompeii. Archaeologists (people who study the past) can recreate the shapes of their bodies. First they fill hollows left by the bodies with wet plaster of Paris and let it harden. Then they remove the cast from the rock in which the bodies fell.

1 Make sure the funnel is dry first. Place it in the top of the bottle and pour in some baking soda. Now pour the vinegar into the funnel from the jug and into the bottle.

2 Remove the funnel. Quickly fit the neck of the balloon over the top of the bottle. Notice that the vinegar and soda are fizzing and giving off bubbles of gas.

3 The balloon starts to blow up because of the pressure, or force, of the gas in the bottle. The more gas given out, the more the balloon fills. Don't burst the balloon!

P R O J E C T

SUFFOCATING GAS

You will need: funnel, bottle, baking soda, vinegar, jug, modelling clay, pencil, long straw, tall and short candles, large jar, matches.

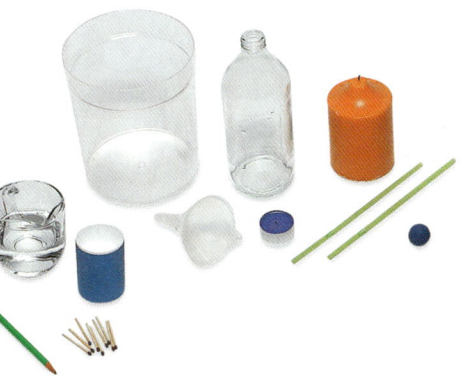

1 Place the funnel in the bottle and add the baking soda. Pour in the vinegar from the jug. This bottle is your gas generator. The gas produced is carbon dioxide.

2 Knead a piece of modelling clay until it is soft, then push it into the mouth of the bottle. Make sure it fits tightly. This will make sure that no gas will escape past it.

Deadly fumes

Clouds of poisonous sulphur fumes billow out from holes on the slopes of Mount Etna, on the Italian island of Sicily. It is one of the most active volcanoes in the world.

3 Make a hole in the clay stopper with the pencil. Carefully push the straw through the hole. Press the clay around the straw.

5 Direct the straw of your gas generator into the bottom of the jar. Keep your arms well away from the candle flames. Soon you will find that the short candle goes out. The carbon dioxide gas has covered it and blocked out the oxygen that would let it burn.

4 Stand both candles in the bottom of the large jar. Ask an adult to light them. Light the short one first to avoid the danger of being burned if the tall candle were lit first.

MOUNT ST HELENS

Picture perfect
Mount St Helens before the May 1980 eruption.

Mount St Helens lies in the Cascade range of mountains near the north-west coast of the USA. This mountain range includes many volcanoes. Before 1980, Mount St Helens had not erupted for 130 years. The mountain began to shake in March 1980. Scientists knew there was about to be an eruption. Many scientists and tourists travelled to photograph what would happen. The progress of the eruption was recorded by hosts of people on the ground, in the air and also by satellite. Nothing prepared the geologists who had gathered there, for the spectacular explosion on the morning of 18 May 1980, however. The blast, the ash clouds, the rain of debris from the volcano, the mud slides, and the poisonous fumes killed 60 people that morning. When the clouds cleared, the mountain had lost 430m in height and acquired a crater 3km across. Mount St Helens was no longer a beautiful piece of tourist scenery.

Blast off
An enormous cloud of thick ash billows from the huge new crater formed when the top of Mount St Helens blew off on 18 May 1980. The cloud rose to a height of more than 20km. It dropped ash over the surrounding region and on towns far away as it blew towards them. In some towns the ash blocked out the Sun. The city of Yakima was particularly badly hit. Over 500,000 tonnes of ash later had to be removed from the area surrounding Mount St Helens.

Before and after
These satellite photographs of the Mount St Helens region were taken before and after the eruption. They show how much land was devastated and covered by ash. The picture on the left was taken a few months before the eruption occurred. The mountain's snow cap is beginning to grow as autumn sets in. The picture on the right was taken about a year after the eruption. Ash covers thousands of hectares of what was once forest land.

Like ninepins

This photograph shows what remained of a forest on the slopes of Mount St Helens after it erupted. Thousands of trees were knocked over by the powerful blast. In places the fallen trees were swept away by an avalanche of rocks, dust and mud, which caused even greater destruction.

Ominous dome

Since the 1980 eruption, Mount St Helens has been quiet. Domes like this near the summit show that magma is still pushing up to the top of the volcano, however. This shows that it is still active.

Blooming again

Only a year after the eruption in 1980, plants are making a comeback on Mount St Helens. Flowers are blooming again on slopes washed clean by rain, and shrubs are pushing their way through the ash.

FACT BOX

• Native Americans of the Pacific Northwest called Mount St Helens *Tah-one-lat-clah* (fire mountain).

• Mount St Helens was previously active between 1832 and 1857.

• The first indication of a forthcoming eruption occurred on 20 March 1980, when an earthquake measuring over 4 on the Richter scale was recorded in the Mount St Helens area.

• On 27 March an explosion rocked the area, caused by an eruption of steam.

• Mount St Helens blew up at precisely 8.32 on the morning of 18 May 1980.

• The crater formed by the eruption measured 3.8km long and 1.9km wide.

Dark as night

Seven hours after Mount St Helens blew, street lighting was needed 140km away in the town of Yakima because the air was filled with black, choking dust.

VOLCANIC ROCKS

The lava that flows out of volcanoes eventually cools, hardens and becomes solid rock. Volcanoes can give off several different kinds of lava that form different kinds of rocks. All these rocks are known as igneous, or fire-formed rocks, because they were born in the fiery heart of volcanoes. They contrast with sedimentary rock, the other main kind of rock found in the Earth's crust. This was formed from layers of silt that built up in ancient rivers and seas. Two of the main kinds of igneous rocks formed by volcanoes are basalt and andesite. Basalt is the rock most often formed from runny lava. This kind of lava pours out of the volcanoes on the ocean ridges and over hot spots. It is dark and dense. Andesite is the rock most often formed from the pasty lava that comes out of the explosive volcanoes on destructive plate boundaries. Because the crystals in both rocks are very small, they are called fine-grained rocks.

Road block
Lava flowing from the Kilauea volcano on Hawaii has cut off one of the island's roads. When the flow stopped, the molten lava had solidified into black volcanic rock with a smooth surface.

Intrusions
Here in Lanzarote, in the Canary Islands, molten rock has intruded, or forced its way through, other rock layers and then hardened. When this happens, geologists call it an intrusion. Volcanic intrusions like this most often occur underground when molten rock forces its way towards the surface. Sheet-like intrusions are known as dykes if they are vertical. If intrusions are horizontal and form between the rock layers, or strata, geologists call them sills.

Obsidian

This volcanic rock is formed when lava cools very quickly. It looks like black glass and is often called volcanic glass.

Basalt

Dark, heavy basalt is one of the most common volcanic rocks. This sample is known as vesicular basalt because it is riddled with holes.

Andesite

Andesite is a lighter-coloured rock than basalt. It is so-called because it is the typical rock found in the Andes Mountains.

Rhyolite

This is another fine-grained rock like basalt and andesite. It is much lighter in colour and weight than the other two, however.

Pumice

Pumice is a very light rock that is full of holes. It forms when lava containing a lot of gas pours out of underwater volcanoes.

Tuff

Tuff is rock formed from the ash ejected in volcanic eruptions. It is fine-grained and quite soft and porous.

The wrinkly skin

The runny type of lava called pahoehoe quickly forms a skin on its surface. This cools first. The lava underneath is still moving and causes the skin to fold and wrinkle. In large flows the surface may cool to form a solid crust, while lava still runs underneath.

Ropy lava

The surface of pahoehoe lava often wrinkles up into shapes that look like bundles of rope. The picture shows this kind of lava, which is known as ropy lava. When a lava flow is thick, it also develops vertical cracks while it is cooling down.

Sandy shores

In most parts of the world, the beaches are covered with pale yellow sand. But in volcanic regions, such as here in Costa Rica, the beaches have black sand. The sand has been formed by the action of the sea beating against dark volcanic rocks and grinding them into tiny particles.

BUBBLES AND INTRUSIONS

Rock slice

A highly magnified picture of a thin slice of the intrusive rock called andesite. When looked at through a microscope, it is possible to see the tiny crystals in this slice of rock.

DISSOLVED GAS

You will need: small jar with tight-fitting lid, bowl, jug, antacid tablets.

In the first project on this page we see how keeping a liquid under pressure stops gas from escaping. The liquid magma in volcanoes usually has a lot of gas dissolved in it. As it rises through the volcano, the pressure drops and the gases start to leak. They help push the magma up and out if the vent is clear. But if the vent is blocked, the gas pressure builds up and eventually causes the volcano to explode. The lava that comes from volcanoes with gassy magma forms rock riddled with vesicles (holes). The pasty lava from some explosive volcanoes sometimes contains so much gas that it forms a light, frothy rock that floats on water. We know this rock as pumice. When rising magma becomes trapped underground, it forces its way into gaps in the rocks and between the rock layers. This process is known as intrusion. The rocks that form when the magma cools and solidifies are called intrusive rocks. Granite is the most common intrusive rock. Often the heat of the intruding magma changes the surrounding rocks. They turn into what are called metamorphic (changed form) rocks, and are the third main rock type, after igneous and sedimentary.

3 Now quickly unscrew the lid from the jar, and see what happens. The whole jar starts fizzing. Removing the lid releases the pressure, and the gas in the liquid bubbles out.

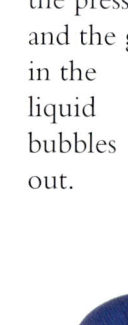

1 Stand the jar in the bowl. Pour cold water into the jar from the jug until it is nearly full to the top. Break up two antacid tablets and drop them into the jar.

2 Quickly screw the lid on the jar. Little bubbles will start to rise from the tablets but will soon stop. Pressure has built up in the jar and prevents any more gas escaping.

P R O J E C T

IGNEOUS INTRUSION

You will need: plastic jar, bradawl (hole punch), pieces of broken tiles, modelling clay, tube of coloured toothpaste.

1 Make a hole in the bottom of the plastic jar with a bradawl, enough to fit the neck of the toothpaste tube in. Keep your steadying hand away from the sharp end of the bradawl.

2 Place the pieces of broken tiles on the bottom of the jar. Keep them as flat as possible. They are meant to represent the layers of rocks we find in the Earth's crust.

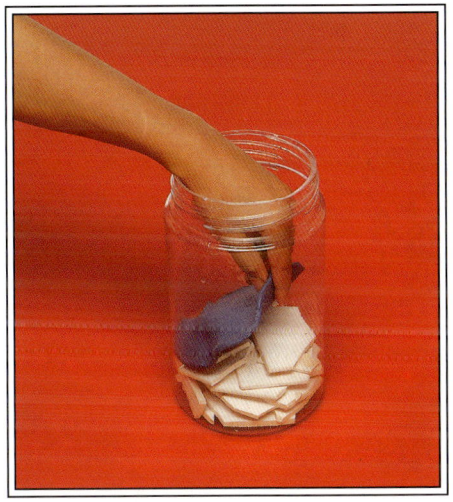

3 Flatten out the modelling clay into a disc as wide as the inside of the jar. Put the disc of modelling clay inside the jar. Push it down firmly on top of the tiles.

5 Squeeze the toothpaste tube. You will see the toothpaste pushing, or intruding, into the tile layers and making the disc on top rise. Molten magma often behaves in the same way. It intrudes into rock layers and makes the Earth's surface bulge.

4 Unscrew the top of the toothpaste tube. and force the neck into the hole you have made in the bottom of the bottle. You may have to widen it a little to get the neck in, but don't make it too wide.

FIERY MINERALS

Some of the rocks that volcanoes produce are very useful to people because they contain valuable minerals. These minerals are a source of metals, such as copper, silver and gold. Minerals are compounds (combinations) of chemical elements that are in the Earth. They form in the intense heat of volcanoes. Rock from volcanoes, such as basalt, looks as though it is a solid black lump. If you look at it under a microscope, however, you can see millions of little specks. These specks are crystals of minerals. The mineral crystals need time to grow. Lava cools so quickly that this does not happen. In volcanic rocks that cool more slowly, crystals have time to grow big enough to see without a microscope. Granite is a rock that forms when magma cools slowly underground. It contains three main kinds of mineral crystals: milky white quartz, pink feldspar, and black mica. Some of the most valuable mineral materials occur in streaks, or veins, in the rocks. They include ores (minerals from which metals such as the lead ore galena can be extracted) and precious metals such as gold.

Quartz crystals
Pure quartz is colourless and is known as rock crystal. It forms hexagonal (six-sided) crystals that end in pyramid shapes. Coloured crystals, such as amethyst, are used as precious stones.

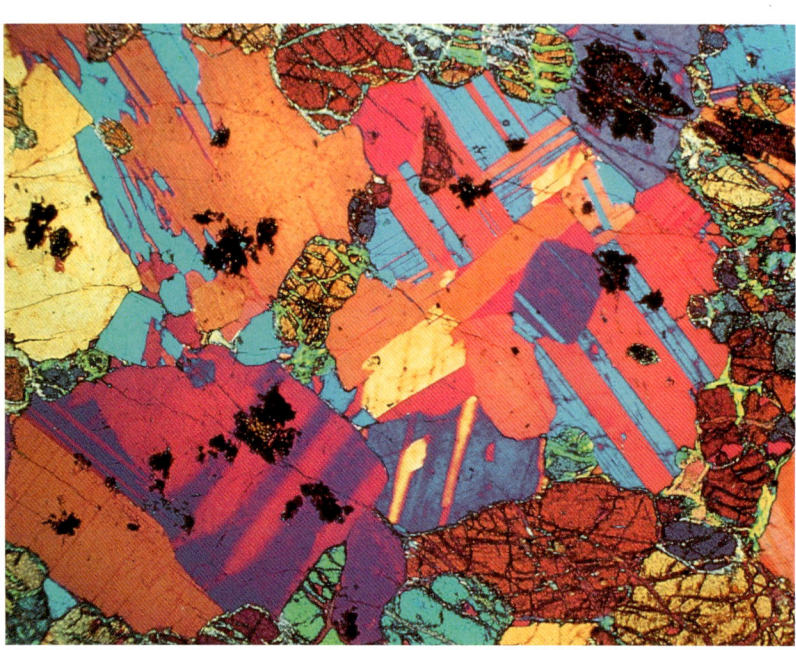

A slice of olivine
The tiny crystals in olivine can be seen here in many colours. Olivine is a mineral often found in volcanic rocks. A special kind of light called polarized light has been shone through a very thin slice of the olivine. Then the slice has been viewed through a special microscope. Transparent, pale green crystals of olivine, known as peridot, are used as gems in the making of jewellery.

Friend or foe?
Most volcanoes give off sulphur when they erupt, and this can collect into huge deposits after a time. Here a worker is cutting sulphur from deposits found around a volcano in Indonesia. Sulphur is very useful in modern industry but it can damage the lungs of those who mine it.

Bomb filling
Volcanic bombs such as these are often blasted out by erupting volcanoes. They are made of cooled lava and may contain other pieces of rock. Inside this bomb a chunk of peridotite was found. Peridotite is a volcanic rock tinged green by olivine crystals.

As nature intended
A diamond pictured in the rock in which it was found. When dug from the Earth, diamonds look rather like dirty glass. Only when they are expertly cut and polished do they display their outstanding beauty and sparkle.

For industry
This collection of diamonds has been cut and polished. They will not be turned into expensive jewellery, however. Their colour is not pure enough and they contain flaws. They are industrial diamonds, which will be used, for example, in drill bits for cutting into rock. Diamond is the hardest of all the minerals on the Earth.

In the vein
The silvery-grey crystals in this rock sample are of the mineral galena, which forms cubic crystals. It is one of the main ores of lead and often contains the metal silver. It is often found with zinc and silver minerals.

Beautiful beryl
Crystals of emerald, one of the finest gems, can be seen buried in a mass of quartz from Colombia, in Central America. Emerald is the green variety of a mineral called beryl. Other coloured varieties of beryl are gemstones too, including aquamarine, a bluish-green mineral.

VOLCANIC LANDSCAPES

Blooms in a cinder desert
Hot cinders once blanketed the ground here. Since then wind-blown seeds have settled, germinated, grown into plants, and flowered. Their roots will help break down the cinders into better soil.

The landscapes in active or recently active volcanic regions are bare and often drably coloured. They do not look as if they could ever be covered with vegetation. However, they are not without beauty, and in time, plants will grow there. The constant action of wind, rain, heat and frost eventually breaks down the newly-formed rocks. The rocks turn into soil, in which wind-blown seeds soon germinate. Providing the climate is suitable, flowers, shrubs, and finally trees will eventually grow again. Sprinklings of ash from further eruptions may, in time, add to and increase the fertility of the soil. But there is always the danger that when a volcano erupts it will destroy all the plants that have grown since the last eruption. This can happen in hours. After the enormous eruption of Mount St Helens in 1980, the blast, hot ash cloud and mud avalanches killed everything within 25km. Less than a month afterwards, however, wildflowers began to grow, and soon insects and small animals began returning.

Dead landscape
A typical volcanic landscape in Iceland looks bleak before regeneration. Volcanic peaks tower in the distance, while in the foreground are bare, black volcanic rocks. Winter is coming, and temperatures are struggling to rise above freezing point. Conditions do not seem favourable for plant life.

Suitable soil
Here in another part of Iceland, the ground is carpeted with low-growing plants. They are growing in the thin soil that now covers a former lava field. The action of the weather and primitive plants such as mosses and lichens have broken down the lava into soil deep enough to support larger plants.

Etna's attractions

Giant cinder cones known as the Silvestri craters are on the southern side of Mount Etna in Sicily. The 3,390m-high mountain erupts frequently. It is one of a string of volcanoes in the Mediterranean region. The others include Stromboli, Vulcano and, most famously, Vesuvius. Geologists estimate that Mount Etna has probably been active for more than 2.5 million years. More than 110 eruptions of the volcano have been recorded since 1500BC. A particularly long eruption in 1992 destroyed much farmland and threatened several villages. The city of Catania, on the lower slopes of the mountain, is often showered with ash.

Rich paddies

Terraces of paddy fields are built on the slopes of hillsides in Bali in Indonesia, south-east Asia. The soil is very fertile because of the ash blasted out by the many volcanoes on the island. Terracing helps increase the amount of farming land and conserves water. In the hot, humid conditions, farmers can grow several crops of rice every year.

Fruit of the vine

Grape vines grow in vineyards on fertile land on the Italian island of Sicily. The land has been fertilized for centuries by the ash from regular eruptions of the island's famous volcano Mount Etna, which looms menacingly on the skyline. Etna is Europe's most active volcano.

Unstoppable

Nothing can stop this thick ribbon of lava, from an erupting volcano named Kimanura, smashing its way through tropical forest in Zaire. There are a number of active volcanoes along Zaire's eastern border. The border is in the Great Rift Valley, where plates meet, causing volcanic activity.

LETTING OFF STEAM

Hothouses
Here in Iceland, the greenhouses are heated by hot water piped in from the many hot springs in the rocks. The people of Iceland rely more on geothermal heating than any other nation.

The molten rock, or magma, in the earth's mantle does not always break out to create volcanoes. Sometimes it stays beneath the Earth's crust. There it causes other volcanic features. They are called geothermal features because they are almost always caused by the Earth (geo) creating heat (thermal) in underground magma that then affects underground water. The most spectacular geothermal feature is the geyser. This is a fountain of steam and water that erupts from holes in the ground. Vents (holes) called fumaroles, where steam escapes gently, are more common. They may also give out carbon dioxide and sulphurous fumes. Also common are hot springs, where water becomes heated in underground rocks to a temperature above body heat (about 37°C). Some hot springs can be twice this hot. Many are rich in minerals. For centuries people have believed that bathing in these mineral-rich springs is good for health.

Iceland *geysir*
A column of steam and water spurts out of the ground and high into the air as Iceland's Strokkur geyser erupts. It is just one of hundreds of geysers found in Iceland. The word geyser comes from the Icelandic word *geysir* (upwards force). Geysers may erupt every few days or hours. Some erupt at such exactly regular intervals that people can set their watches by them.

Boiling tar
In a volcanically active region in New Zealand, geothermal heating is causing this tarpit to boil and bubble. New Zealand was one of the first countries to tap geothermal energy for power production.

Hot dip

Icelanders enjoy the pleasures of a hot spring. There are hundreds of hot springs dotted around the island. The water from some of them is piped into towns to provide a cheap form of central heating for public buildings and homes. Geothermal heating has many advantages over conventional systems. It does not cause any pollution and is renewable. Geothermal energy will be available until the Earth cools in a billion years from now.

Yellowstone springs

A hot spring makes a colourful sight in Yellowstone National Park in Wyoming, USA. The intense blue of the clear water contrasts with the yellow and orange minerals that have been deposited by evaporation around the edges. Yellowstone is the foremost geothermal region in the USA.

Gleaming terraces

Looking like a frozen waterfall, white terraces of travertine are found in many hot-spring regions, as here in Yellowstone National Park, USA. Travertine is made up of the mineral calcite.

Steam power

A geothermal power station in Iceland makes use of natural geyser activity. Steam is piped up from underground and fed to turbogenerators to produce electricity.

FACT BOX

• One of the most famous geysers in the world is Old Faithful, in Yellowstone National Park in Wyoming, USA. This geyser erupts regularly about once every 45 minutes.

• Yellowstone National Park also boasts the tallest geyser in the world. Known as Steamboat, its spouting column has been known to reach a height of more than 115m.

GEYSERS AND MUDLARKS

All the different kinds of thermal (heat) activity that go on in volcanic regions have the same basic cause. Water on the Earth's surface trickles down through holes and cracks into underground rocks that have been heated by hot magma far below. The water becomes superheated to temperatures far above boiling point (100°C). It does not boil, however, because it is under huge pressure. Eventually, this very hot water may turn to steam and escape from a fumarole (vent where steam escapes). The hot water can also mix with cooler water to create a hot spring, or with mud to form a bubbling mud hole. Sometimes it turns into steam at the bottom of a column of water, creating a steam explosion that blasts water out of the ground as a geyser. The first project shows you how to make a geyser using air pressure to force out water. Blowing into the top of the bottle increases the air pressure there. This forces the coloured water out of the bottle through the long straw.

Waterspout
Superheated steam and water spout high into the air from the Lady Knox geyser at Waiotapu, in New Zealand. A cone of minerals has built up around the mouth of the geyser, which usually erupts for about an hour.

GEYSER ERUPTION

You will need: modelling clay, long bendy straws, jug, food colouring, large plastic bottle, large jar.

1 Make two holes in a little ball of clay and push two bendy straws through it as shown in the picture. Push another straw through the end of one of the first two straws.

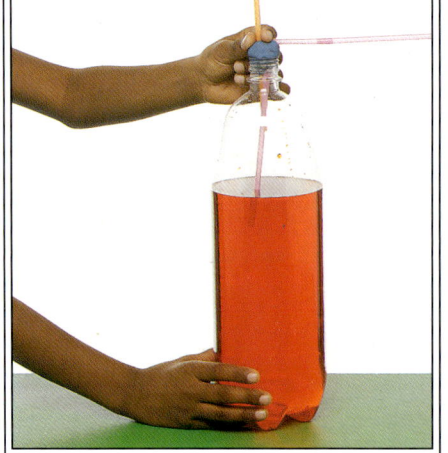

2 Pour water into the jug and add the colouring. Then pour it into the bottle. Push the clay stopper into the neck so that the lengthened straw dips into the coloured water.

3 Place the jar under the other end of the lengthened straw and blow into the other straw. Water spurts out into the jar. If the long straw was upright, the water would spout upward like a geyser.

MUDBATHS

You will need: cornflour, chocolate powder, measuring jug, mixing bowl, wooden spoon, milk, saucepan, oven mitt.

1 Mix together two tablespoons of cornflour and two of chocolate powder in the bowl, using the spoon. Stir the mixture thoroughly until it is an even colour.

2 Pour about 300ml of milk into the saucepan, and heat it slowly on a hotplate. Keep the hotplate on a low setting to make sure the milk does not boil. Do not leave unattended.

4 Pour the creamy mixture into the hot milk in the saucepan, still keeping the hotplate on a low heat. Holding the handle of the saucepan with the oven mitt, stir constantly to prevent the thick liquid sticking to the bottom of the saucepan.

3 Add some cold milk, little by little, to the mixture of cornflour and chocolate in the bowl. Stir vigorously until the mixture has become a thick smooth cream.

6 Soon your hot liquid mud will start sending up thick bubbles, which will burst with gentle plopping sounds. This is exactly what happens in hot mud pools in volcanic areas.

5 If you have prepared your flour and chocolate mixture well, you will now have a smooth hot liquid looking something like liquid mud.

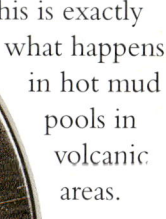

CHANGING CLIMATES

Mighty blast
In August 1883, the volcano Krakatoa blasted itself apart. The ash clouds from the volcano rose high into the atmosphere, spreading out and travelling in a band around the world.

Volcanoes can have a noticeable effect on the weather locally (nearby) when they erupt. Over weeks or months, they can affect climates around the world. Locally, volcanoes can set off lightning flashes. These break out when static electricity builds up in the volcano's billowing ash clouds and then discharges like a gigantic electric spark. The ash clouds from volcanoes may be so thick that they block out the sunlight and turn day into night. This happened for hours in the region around Mount St Helens after the eruption in 1980. It also happened for days during the eruption of Mount Pinatubo in the Philippines in 1991. The Mount Pinatubo eruption also had longer-term effects. The gas and dust it gave out stayed in the atmosphere (air) for months, producing spectacular sunsets. So much escaped into the high atmosphere that it cut down the sunlight reaching the ground. This cooled down the Earth's climate enough to affect weather patterns for a number of years. The Mexican volcano El Chichon, which erupted in 1982, had the same effect. Its ash had a particularly high sulphur content. Chemicals containing sulphur are believed to block sunlight most.

A dying breed
The fossil skeleton of a pterosaur, a flying dinosaur that became extinct (died out) about 65 million years ago. It might have perished as a result of the Earth being plunged into darkness after planetwide volcanic eruptions.

Blowing its top
The crater at the top of the Mexican volcano El Chichon. Until April 1982 it had a jungle-covered conical summit. But on 4 April, this was blasted away in an explosion that drove ash high into the atmosphere.

Fissure eruptions
This series of volcanic cones follows a long fault in Iceland called the Skaftar fissure. Massive ash eruptions occurred along the fissure in 1783 and caused cold winters in Europe.

Chilly winters
The ash and gases from volcanoes can stay in the atmosphere for years. If enough volcanoes erupted at the same time, winters could be much colder than usual. In very cold winters in the 1800s, people held markets called frost fairs on deep-frozen rivers.

Red night delights
Spectacular sunsets often occur when eruptions throw dust and ash into the air. In 1991, before Australians saw sunsets like this following the eruption of Mount Pinatubo in the Philippines. Mud slides that followed the eruption killed more than 400 people.

Astronauts' eye view
Space shuttle astronauts took this picture of the ash cloud rising from the eruption of the volcano at Rabaul, New Guinea in 1994. It was estimated that the cloud rose between 20 and 30km into the sky. On the ground ash fell 75cm deep and destroyed two-thirds of the town of Rabaul.

FACT BOX

• The Indonesian volcano Tambora, which erupted in 1815, produced so much ash that world temperatures fell sharply in the following year. New England, in the eastern USA, had severe frosts in August.

• Mount Pinatubo, which erupted in the Philippines in June 1991, released nearly 8 cubic km of ash. This totals eight times as much ash as at the eruption of Mount St Helens.

OUT OF THIS WORLD

Earth is not the only place in the universe that has volcanic activity. Many other planets and moons in our Solar System have had volcanoes erupting on their surface at some time in their history. Two of the planets nearest to us, Venus and Mars, were affected by volcanoes. Venus and Mars are both terrestrial (Earth-like) planets, with a similar rocky structure to Earth. The whole landscape of Venus, revealed by the Magellan radar probe between 1990 and 1994, is volcanic. There are volcanoes everywhere. Most of the surface consists of vast lava plains stretching for thousands of kilometres. Mars has fewer volcanoes, but they are gigantic. The record-breaker is Olympus Mons, which is more than five times the height of Earth's highest mountain, Mount Everest. Nearer home, volcanoes have been a major force in shaping our Moon. The dark patches we see on the Moon at night are flat plains that flooded with lava when massive volcanic eruptions took place long ago. But some of the most interesting volcanoes lie much farther away, on one of Jupiter's moons, Io. Its volcanoes pour out liquid sulphur.

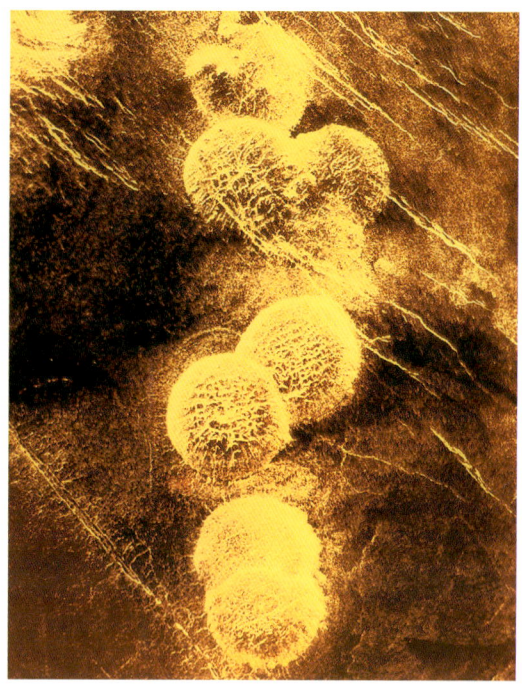

Volcanic pancakes
A series of volcanic features on Venus are called pancake domes. Scientists think they form when molten rock pours out of flat ground, spreads out and hardens.

The greatest
This is the biggest volcano on Mars, and one of the biggest we know in the whole Solar System. It is named Olympus Mons, or Mount Olympus. The volcano is 600km across at the base, and it rises to a height of some 27km.

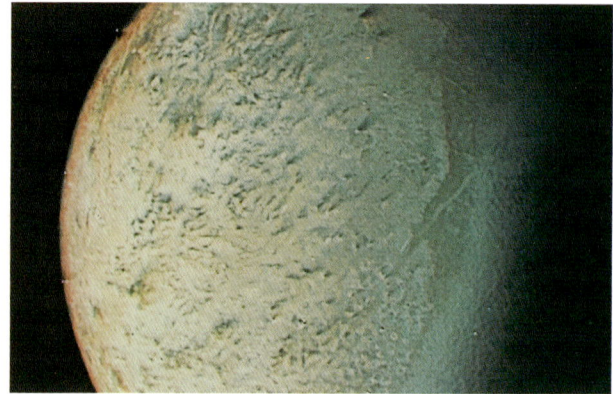

Triton's eruptions
Volcanic eruptions take place on Triton, the largest moon of Neptune. Because the moon is very cold (about −235°C), its volcanoes give off liquid nitrogen. Dark material comes out as well, causing the dark streaks visible in the picture.

Out on a limb

A volcano erupts on the edge, or limb, of Jupiter's moon Io. It shoots gas and dust hundreds of kilometres into space as well as pouring molten material over Io's surface. The material that comes out of the volcano is not molten rock, however. It is a liquid form of the chemical sulphur. Sulphur is a yellow-orange colour, which explains why Io is such a colourful moon. Io's volcanoes were among the many astonishing discoveries made by the Voyager space probes. They visited the outer planets between 1979 and 1989.

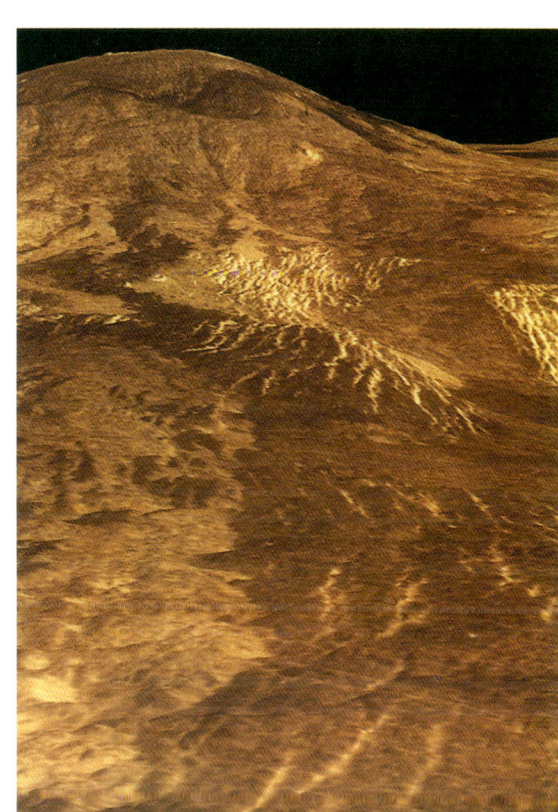

Volcano on Venus

One of Venus's many volcanoes recorded by the Magellan space probe. It has the typical broad dome shape of the shield volcanoes on Earth. Most volcanoes on Venus are of this type and, like all shield-type volcanoes, they pour out runny lava. Repeated eruptions over millions of years have sent rivers of lava streaming for hundreds of kilometres around. Most of the landscape of Venus consists of rolling plains made up of such lava flows. Venus's biggest volcanoes are up to 500km across and several kilometres high, but most are much smaller. Venus has many other volcanic features, including strange, spidery structures called arachnoids.

The lunar seas

The seas on the Moon are flat plains. They were created billions of years ago when lava flooded into huge craters made by meteorites. The picture shows part of the Moon's largest sea, which is called the Ocean of Storms. This sea covers more than 5 million sq km, and is more than half as big again as the Mediterranean Sea on Earth. The large crater in the picture is called Kepler by astronomers. It is about 35km from one side to the other.

THE QUAKING EARTH

Many people consider the city of San Francisco, in California, to be one of the most beautiful in the world. It has a stunning setting on the USA's west coast and enjoys a pleasant climate. But living in the city has one major disadvantage. San Francisco sits nearly on top of a line of weakness in the Earth's crust known as the San Andreas fault. The fault marks the boundary between two of the plates in the Earth's crust, the eastern Pacific and the North American plates. These plates are trying to slide past each other. They do this jerkily and when they do, the ground shakes violently. Earthquakes occur around the boundaries of all the plates on the Earth's surface, especially where the plates are colliding. This is why they often occur in the same places as volcanoes, which also occur at plate boundaries. Tens of thousands of earthquakes take place every year throughout the world, but only about 1,000 of them are powerful enough to cause damage. Such earthquakes are incredibly destructive. Most only last for a few seconds, but in that short time they can reduce whole cities to rubble and kill thousands of people. The main earthquake is always followed by smaller ones. These are called aftershocks and happen when the rocks along the edge of the fault settle into their new positions. These aftershocks can also cause a lot of damage.

Famous fault
The most famous earthquake-producing fault in the world is the San Andreas in California, USA. It runs for hundreds of kilometres, passing close to the cities of Los Angeles and San Francisco.

Not so grand
This old print shows the chaos and destruction that earthquakes can bring. This earthquake was in 1843, in the port of Pointe-à-Pitre on the island of Grande Terre. It is one of the Guadeloupe group of islands in the Caribbean.

Housing slump

An earthquake in San Francisco in October 1989 caused whole rows of houses to collapse or damaged them beyond repair. In only a few seconds, more than 60 people were killed.

> ### FACT BOX
>
> • The powerful earthquake that hit San Francisco on 18 April 1906 and the fire afterward destroyed the whole city. Almost 700 people died.
>
> • It is estimated that as many as 750,000 people were killed in the Chinese city of Tangshan and surrounding regions by the earthquake there on 28 July 1976.

Anchorage in ruins

In March 1964, a powerful earthquake hit Anchorage, in Alaska. It was one of the longest ever recorded. The town and surrounding regions shook for four long minutes. Roads disappeared into the ground.

One-way street

An earthquake demolished one side of the Kalapana road in Hawaii in 1984. The ground was set shaking when the volcano Kilauea rumbled into life. Earthquakes occur frequently in volcanic regions.

No highway

In a 1994 earthquake that rocked Los Angeles, an elevated section of highway was shaken off its supporting piers (legs). Elevated roads are difficult to make earthquake-proof. The piers they stand on shake easily in earthquakes.

Kobe's killer waves

Some of the destruction caused by the powerful earthquake that struck the city of Kobe, Japan, in 1995. Multistorey apartment blocks collapsed like packs of cards. It was the country's most destructive earthquake since 1923.

SLIPS AND FAULTS

Every earthquake, from the slightest tremor you can hardly feel, to the violent shaking that destroys buildings, has the same basic cause. Two blocks of rock grind past each other along a fault line where the Earth has fractured (the crust has split). There are several kinds of fault. At the San Andreas fault in California, the blocks are sliding past each other horizontally. This is called a transform fault, or strike-slip fault. In a normal fault, the rocks are pulling apart and one block slides down the other. In a thrust fault, the blocks are pressing together, causing one to ride up above the other. Because the edges of the blocks in contact at a fault are very uneven, friction (resistance to movement) locks them together. As they try to move, the rocks become strained and stretched. In the end, the strain in the rocks grows so great that it overcomes the friction. The two blocks suddenly move apart. The energy in the rocks is released as earthquake waves that cause great destruction.

All shook up
Traffic lights fell over in Los Angeles, USA, after an earthquake in January 1994 which killed 60 people.

FAULT MOVEMENTS

You will need: two wooden blocks, jar of baby oil, drawing pins, sheets of sandpaper.

1 Hold a block in each hand so that the sides of the blocks are touching. Pushing gently, try to make the blocks slide past each other. You will find this quite easy.

2 Wet the sides of the blocks with the oil, and try to slide them again. You should find that it is easier. The oil has lessened the friction between the blocks.

3 Pin sheets of sandpaper on the sides of the blocks, and try to make them slide now. You will find it much more difficult. The sandpaper is rough and increases friction between the blocks.

P R O J E C T

QUAKES

You will need: scissors, strong elastic band, ruler, plastic seed tray (without holes), piece of card, salt.

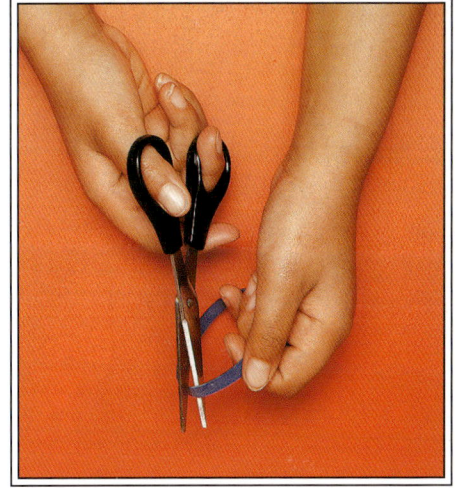

1 With the scissors, cut the elastic band at one end to make a long strip. This represents a layer of rock inside the Earth before it is affected by an earthquake.

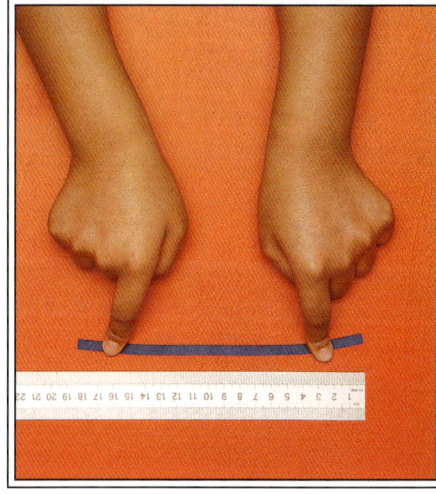

2 Measure the strip of elastic with a ruler. This represents the original length of the rock in the ground. Make a note of how long the elastic is at this stage.

3 Stretch the elastic band and hold it over the tray. Rocks get stretched by pulling forces inside the Earth during an earthquake.

4 Ask a friend to hold the card on top of the elastic and sprinkle some salt on it. The salt layer now on the card represents the surface of the ground above the stretched rock layer.

5 Now let go of the ends of the elastic. Notice how the salt grains on the card are thrown about. This was caused by the energy released when the elastic shrunk.

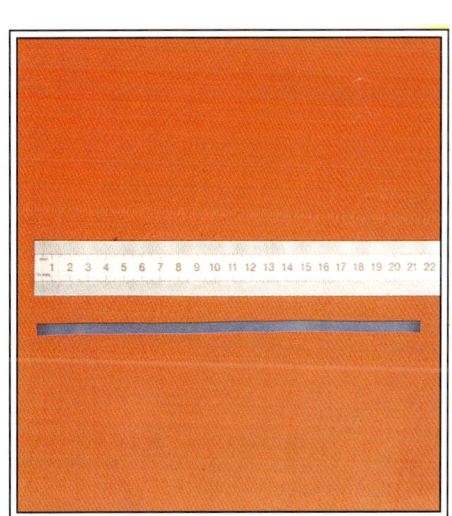

6 Finally, measure the strip of elastic again. You will find that it is slightly longer than it was at the start. Rocks are often permanently stretched a little after an earthquake.

No highway

Part of the elevated highway in Kobe that collapsed during the 1995 earthquake. The supporting columns were shaken into pieces by the force of the tremors.

TREMENDOUS TREMORS

The movement of rocks that causes earthquakes usually occurs deep inside the Earth's crust. The exact point where the rocks start to break, or fracture, is known as the focus. This can lie as deep as hundreds of kilometres or as close as a few tens of kilometres. At the surface, the most violent disturbance occurs at a point directly above the focus, called the epicentre. The closer the focus, the more destructive is the earthquake. The earthquake that struck Kobe, Japan, in 1995 was so destructive because its focus was only about 15km deep. The focus of the great Alaskan earthquake of 1964 was not much deeper and caused massive destruction. The epicentre of that earthquake was on the coast of the Gulf of Alaska, and also caused the seabed to rise. This created a surge of water up to 21m high – it was a tidal wave, or tsunami. The tsunami devastated coastal towns and islands for hundreds of kilometres around.

Earthquake-proof
The Transamerica building in San Francisco is very distinctive. It has been built with flexible foundations. These should allow it to withstand the shaking that earthquakes bring.

Struts hold pipeline above ground

Flexible pipe
The Transalaska Pipeline snakes through the wilderness of Alaska, carrying oil south from the oilfields of the North Slope. It is built above ground. There are zigzags in places to allow it to move if and when earthquakes occur. The pipeline stretches for some 1,285km.

Epicentre

Fault line

Cracked rocks

Focus

Earthquakes in focus
Most earthquakes originate in rock layers many kilometres below the surface, at the focus. The most intense vibrations on the surface are felt immediately above the focus, at the epicentre.

Fire alarm

Fire breaks out in a gas main following a minor earthquake in Los Angeles. Underground pipelines carrying gas, oil or water are damaged easily when the ground vibrates. They can cause additional hazards to victims of the earthquake and their rescuers. The pipes break and their contents leak. Gas and oil catch fire easily. Water pipes can cause large floods.

Wall of water

This old print shows the tsunami (tidal wave) that followed the explosion of the volcano Krakatoa in 1883. Most of the people who died as a result of the eruption were drowned when this wall of water swept across the neighbouring low-lying islands.

Shipwrecked

One of the many fishing boats that were wrecked by the tidal wave that followed the powerful earthquake in Alaska in 1964. The wave devastated all the coastal communities around the Gulf of Alaska.

Displaced persons

Tents provide temporary shelter for the inhabitants of a town in nothern Turkey, following the earthquake in 1999. It is not yet safe to return home, even to houses that suffered little damage. A main earthquake is always followed by a number of aftershocks. If these are strong enough, they may bring down even more buildings.

The safest?

The designer of this odd-shaped building in Berkeley, California, boasts that it is the world's safest building. He claims that it can withstand the most powerful earthquakes. Berkeley is not far from the notorious San Andreas fault. It may not be long before we find out whether he is right or wrong about how safe his house is.

MAKING WAVES

Ripples in the street
The waves that travel through the surface rocks make the ground ripple. Afterwards, the ripples can often be seen. This road has been affected by waves in the ground. Now the surface of the road is like a wave.

The enormous energy released by an earthquake travels through the ground in the form of waves. Some waves are rather like water waves. They can literally make the ground ripple up and down. Others make the ground shake from side to side, which makes them very destructive. Waves also travel deep underground from an earthquake. The primary (P) waves travel fastest. They travel through rocks in the same way that sound travels through the air, as a series of pressure surges (pushing motions). The secondary (S) waves are slower than the P waves. They travel up and down and from side to side. They are rather like the wave you can see in a rope when you shake it up and down.

NEWTON'S CRADLE

You will need: large beads, lengths of wool, sticky tape, cane, four wooden blocks.

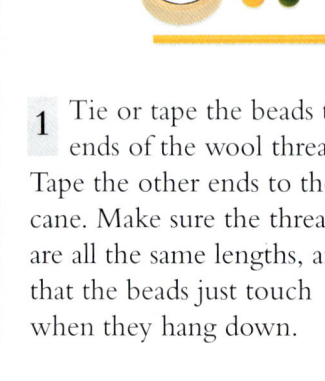

1 Tie or tape the beads to the ends of the wool threads. Tape the other ends to the cane. Make sure the threads are all the same lengths, and that the beads just touch when they hang down.

2 Prop up the cane at both ends on a pair of blocks supported by more blocks underneath. The blocks should be high enough to stop the beads from touching the table. Secure the ends with tape. Lift up the bead at one end of the row and let go. Look what happens to the other beads.

3 The beads in the middle do not move, but the one at the other end flies up. The energy of the falling bead at one end travels as a pressure wave through the middle ones. Then it reaches the bead at the other end and pushes it away.

PROJECT

TREMORS

You will need: set of dominoes, card.

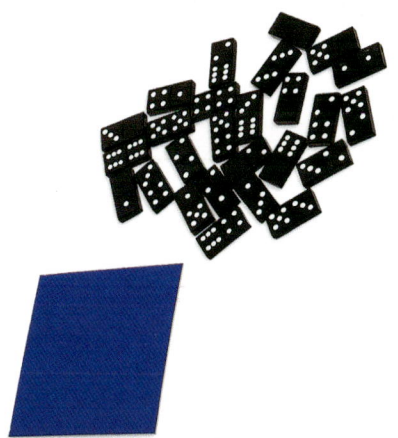

1 This project investigates how the energy in waves varies with distance. Near the end of a table, build a simple house out of dominoes. Stand them up on edge.

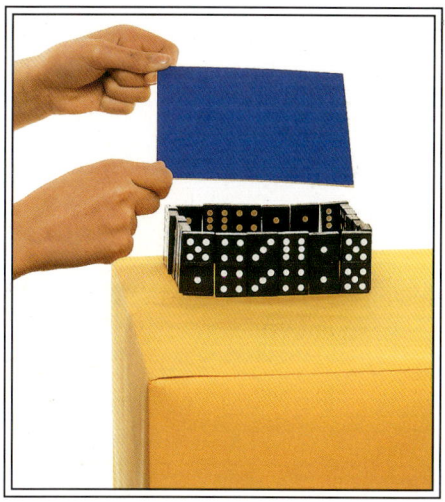

2 Place the card on the dominoes to make the roof of your house. Many people in earthquake zones live in the simplest of houses, built not too differently from this one.

3 Go to the opposite end of the table and hit it with your hand, but not too hard. What happens to your domino house? Probably it shakes, but still stays standing.

Leaning tower blocks
After a major earthquake, buildings lean at all angles as the shock waves destroy their foundations. The 1995 Kobe earthquake in Japan damaged nearly 200,000 buildings.

4 Now go back to the other end where your house is, and hit the table again with the same amount of force. What happens to your domino house this time?

5 Your house comes tumbling down. The waves you create when you hit the table are strong enough to knock down the house when it is nearby. When you hit the table from the opposite end, which is further away from the dominoes, the waves weaken as they travel. They are too weak to knock down the house by the time they reach it.

SEISMIC SCIENCE

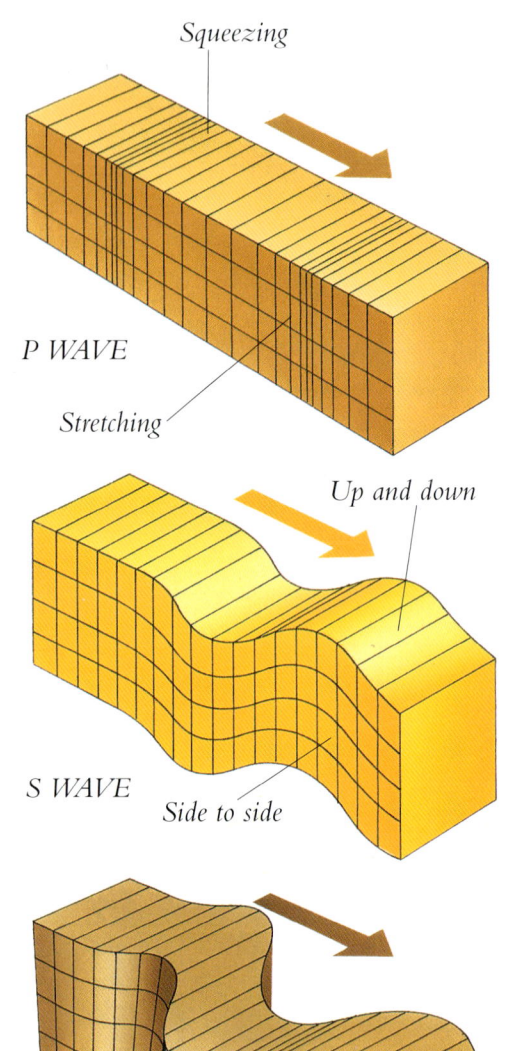

Squeezing

P WAVE

Stretching

Up and down

S WAVE

Side to side

LOVE WAVE *Side to side*

RAYLEIGH WAVE

Rippling up and down

Geologists who specialize in the study of earthquakes are called seismologists. These scientists call the waves that earthquakes create in the rocks seismic waves. The main instrument they use to detect and measure earthquakes is called a seismograph. Modern seismographs record the tremors (waves) an earthquake creates as a readout on a screen. The trace is known as a seismogram. It shows clearly the different waves earthquakes produce. The primary (P) waves, arrive first because they usually travel at speeds of more than 20,000km/h. The secondary (S) waves arrive next. They usually travel at only about half the speed of the P waves. Finally come the surface waves. Among other things, seismologists can tell from a seismograph how strong an earthquake is. The strength, or magnitude, of an earthquake is usually measured on the Richter scale, invented by Charles Richter. Other scales are also used, however, particularly one named after the Italian volcanologist Giuseppe Mercalli.

Making waves
This illustration shows four different ways in which earthquake waves travel through the ground. The primary (P) wave is a compression (squeezing) wave. It compresses, then stretches, rocks it passes through. The secondary (S) wave produces a side-to-side, shaking action. Love waves travel on the surface, making the ground move from side to side. Rayleigh waves are surface waves that move up and down. These two waves are named after the scientists who were the first to study them closely.

Charles Richter (1900–1985)
Charles F. Richter was a US seismologist. In 1931 he worked out a scale for measuring the relative strengths, or magnitudes, of earthquakes, based on the examination of seismograms.

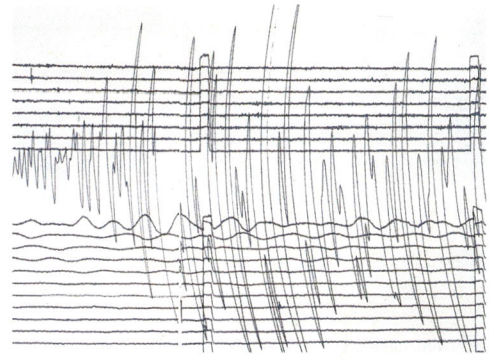

On Vesuvius
An Italian seismologist looks at an old seismograph at the observatory on Mount Vesuvius, near Naples. The building dates from 1845.

Bad vibrations
This is a seismogram of a moderate earthquake in California, USA, in 1989. The widest vibrations show the strongest earth tremors.

Looking for moonquakes
Apollo 11 astronaut Edwin Aldrin sets up instruments on the Moon in 1969. One was a seismometer, designed to measure moonquakes, or ground tremors on the Moon. Seismometers were set up at the other *Apollo* landing sites. They helped scientists work out the structure of the Moon.

Vibrating needle
A close-up picture shows the needle and drum of a seismograph. These machines are being replaced by electronic ones. They will be linked to computers that are able to record waves from earthquakes digitally.

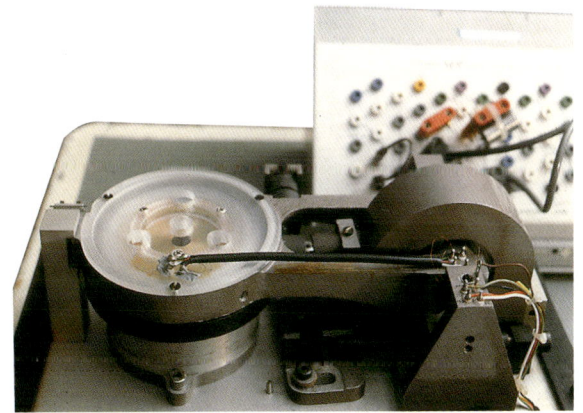

Round and round
A seismologist changes the paper roll on a seismograph at an Antarctic research station. There are many scientific observation stations in the Antarctic. People stay in them for months studying the Earth and weather.

FACT BOX

• The work in the early 1900s of an Eastern European meteorologist (weather scientist) named Andrija Mohorovicic led to the discovery of the layered structure of the Earth.

• The United States National Earthquake Information Service is one of the key seismic centres in the world. It records around 60,000 seismic readings every month.

BUILDING SEISMOGRAPHS

There are thousands of seismic centres scattered around the world. Within minutes of a quake, seismologists in different countries are analysing the seismograms from their seismographs. Then they will compare notes with scientists in other countries and will be able to pinpoint the epicentre and focus of the quake, its strength and how long it lasted. The Italian scientist Luigi Palmieri built the first seismograph in 1856. All seismographs work on the same principle. They use a heavy weight supported by a spring inside a frame. When an earthquake occurs, it shakes the instrument. The heavy weight tends to stay where it is because of its inertia (resistance to change). A pen attached to the weight records the shaking movement as a wavy line drawn on paper wrapped round a rotating drum. The same principle of the inertia of a heavy weight is used to detect tremors in the do-it-yourself seismograph shown in the project here.

Out of a dragon's mouth
This is a model of a seismoscope built by a Chinese scientist of the past called Zhang Heng. The movement of an earthquake shakes a ball out of a dragon's mouth and into a toad's mouth below.

BUILDING A SEISMOGRAPH

You will need: cardboard box, bradawl (hole punch), sticky tape, modelling clay, pencil, felt-tip pen, string, piece of card.

1 The cardboard box will become the frame of your seismograph. It needs to be made of quite stiff card. The open part of the box will be the front of your instrument.

2 Make a hole in what will be the top of the frame with the bradawl (hole punch). If the box feels flimsy, strengthen it by taping round the corners as shown in the picture.

3 Roll a piece of clay into a ball and make a hole in it with the pencil. Now push the felt-tip pen through the clay so that it extends a little way beyond the hole.

P R O J E C T

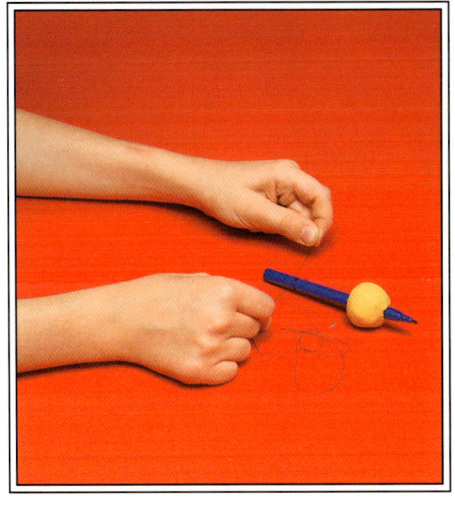

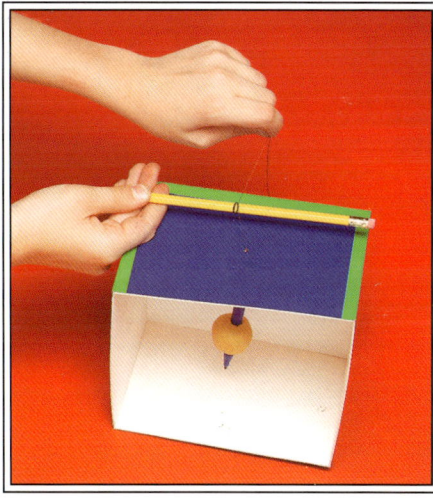

4 The pen and clay bob will be the pointer of your seismograph and make a record of earthquake vibrations. Tie one end of the piece of string to the top of the pen.

5 Thread the other end of the string through the hole in the top of the box. Now stand the box upright and pull the string through until the pen hangs free.

6 Tie the top end of the string to the pencil and roll the pencil to take up the slack. When the pen is at the right height (just touching the bottom) tape the pencil into position.

9 You do not have to wait for an earthquake to test your seismograph. Just shake or tilt the frame. The suspended pen does not move but it marks the piece of card, giving you your very own seismogram.

7 Place the card in the bottom of the box under the pen. If you have adjusted it properly the tip of the pen should just touch and mark the card.

8 Your seismograph is now complete and ready for use. It uses the same principle as a proper seismograph. The heavy bob, or pendulum, will be less affected by shaking motions than the frame.

FIELDWORK

The scientists who study volcanoes and earthquakes spend a great deal of time in the field (on the spot) around active volcanoes and in earthquake zones (places where earthquakes commonly take place). Volcanologists keep an eye on many active volcanoes all the time, looking for any changes that may signal a new eruption. Permanent observatories have been built on volcanoes near centres of population, such as Mount Vesuvius and Mount Etna in Italy. Hopefully the volcanologists can give advance warnings to people who could be at risk. When eruptions do take place, they chart the direction of lava flows and take temperatures and samples of lava and gases. The thermometers they use are not the mercury–in–glass kind. Those would melt at 1,000°C and at the greater temperatures found around volcano sites. Volcanologists use thermocouples to measure temperatures. These are made of metals. Seismologists spend their time in earthquake regions setting up and checking instruments that can record ground movements. This is part of their study to try and predict earthquakes.

Studying creep
A scientist measures movements along a fault using a creepmeter. The two parts of the creepmeter are on either side of the fault line.

Watching Etna
Mount Etna, in Italy, is the highest volcano in Europe. It has been erupting for more than 2.5 million years. Shown above is one of the three observatories set up on its slopes in the mid 1800s. The town of Catania lies on the slopes of Mount Etna. The volcano is watched constantly because if it erupts it could destroy Catania and other villages nearby. Fortunately eruptions on Etna happen quite slowly.

Hard hat job
A geologist checks a lava flow from Hawaii's highly active volcano Kilauea. The sides have already cooled and solidified, which helps shield him from the heat given out by the river of molten rock beneath. He wears a hard hat to protect himself from falling debris.

FACT BOX

• China is particularly earthquake-conscious and has more than 250 seismic stations and 5,000 observation points.

• Some animals behave oddly before earthquakes. Before the Kobe earthquake of 1995 in Japan, vast shoals of fish were seen swimming on the surface of the sea nearby. Flocks of birds were seen flying in crazy patterns. The earthquake struck only hours later.

Space age suits

These volcanologists look quite similar to astronauts in their protective suits. They are carrying out research in Ethiopia about the lava lake at Erta Ale volcano. The suits they wear have a shiny silvery coating. This reflects the heat from the hot lava away from their bodies and so helps to keep them cool. The researchers also wear protective helmets to shield their faces, particularly their eyes, from the heat. Erta Ale has been erupting since 1967.

Laser checking

Seismologists sometimes use a space satellite called Lageos to check for ground movements. Two identical lasers are set up, one on each side of a fault. Movement in the ground affects the time it takes for the laser beams to go to and from the satellite. These time differences tell the scientists the ground is moving.

On the ice

Geologists carry out seismic surveys in Antarctica to study the rock layers under the ice. In places the ice is more than 4,000m thick. Mount Erebus is the only volcano on the continent.

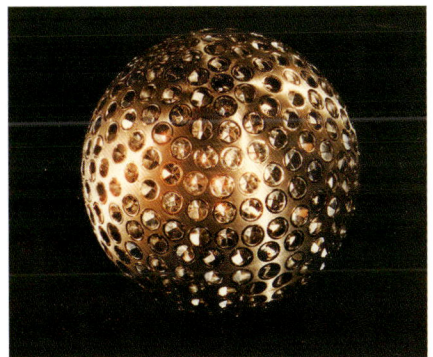

Lageos satellite

Thumping good idea

In the past, seismologists set off explosives to send shock waves they could measure through the rocks. Nowadays they mostly use special vibrator trucks, which thump the ground to create waves.

PROJECT

MEASURING MOVEMENTS

The seismograph is the most important instrument for seismologists once an earthquake has happened. But these scientists use many other instruments, in particular to detect how the ground moves in areas where earthquakes might occur. The San Andreas fault in California is criss-crossed with seismic ground stations, some using laser beams and other electronic devices and others with relatively simple instruments. An extensometer measures stretching movements in the rocks. A magnetometer detects minute changes in the Earth's magnetism that often occurs when rocks move. A creepmeter measures movements along faults. Our two projects show how to make simple versions of instruments called the gravimeter and the tiltmeter. The gravimeter measures slight changes in gravity. When changes occur, the pull on a heavy mass changes, which will make a mass and a pointer attached to it move over a scale. The tiltmeter detects whether rock layers are tilting by comparing the water levels in two connected containers.

Seismic survey
Seismic researchers carry out an accurate survey of the ground in an earthquake region. By comparing their readings with past records, they can tell if any ground movements have taken place.

GRAVIMETER

You will need: strip of sticky paper, pen, large jar, modelling clay, elastic band, toothpick, pencil.

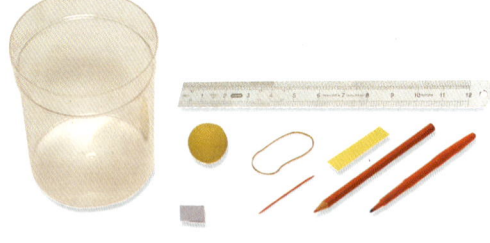

1 Draw a scale on a strip of sticky paper using a ruler and pen. Stick the scale on the jar. In a real instrument this would measure slight changes in gravity.

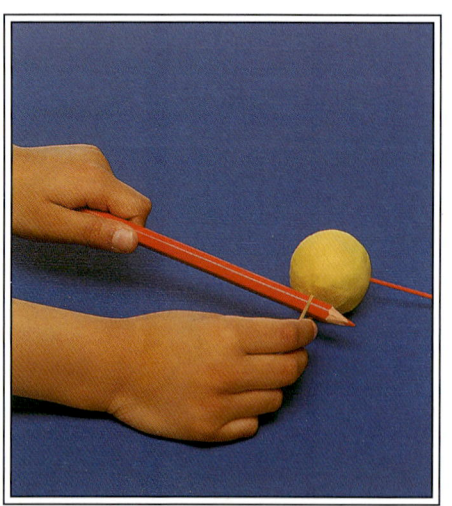

2 Bury one end of an elastic band in a ball of clay. Stick in a toothpick at right angles to the band to act as a pointer. Pass the pencil through the loop of the band.

3 Lower the ball into the jar, dangling from the pencil, so that the tip of the pointer is close to the scale. Rest the pencil on the top of the jar and use bits of clay to stop it moving. If you move the jar up or down, the pointer moves down and up the scale.

P R O J E C T

TILTMETER

You will need: bradawl (hole punch), two transparent plastic cups, transparent plastic tubing, modelling clay, pen, sticky paper, wooden board, adhesive, food colouring, jug.

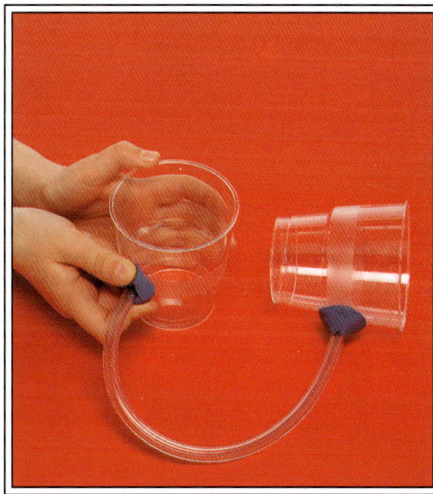

1 Use the bradawl (hole punch) to make a hole in the sides of each plastic cup, just about half-way down. Be careful not to prick your fingers. Ask an adult to help you if you prefer.

2 Push one end of the tubing into the hole in one of the cups. Seal it tight with modelling clay. Put the other end in the hole in the other cup and seal it also.

3 Using the pen, draw identical scales on two strips of the sticky paper. Use a ruler and mark regular spaces. Stick the scales at the same height on the side of the cups.

4 Stick the cups to the wooden baseboard with safe adhesive. Position them so that the tube between is pulled straight, but make sure it doesn't pull out.

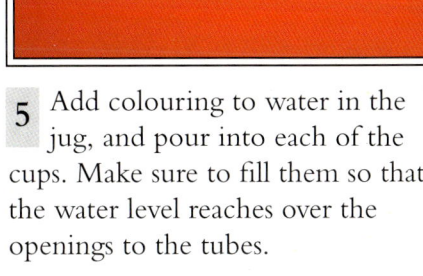

5 Add colouring to water in the jug, and pour into each of the cups. Make sure to fill them so that the water level reaches over the openings to the tubes.

Tilt makes water flow out of upper cup

6 Your tiltmeter is now ready for use. When it is level, the water levels in the cups are the same. When it tilts, the water levels change as water runs through the tube from one cup to the other.

Tilt makes water flow down into lower cup

TO THE RESCUE

When volcanoes erupt and earthquakes strike, they can unleash destructive power equal to hundreds of atomic bombs. The most destructive volcanoes explode and cause ash and mud slides that sweep away everything in their path. Most people caught by these stand no chance and are dead by the time any rescuers can arrive. Earthquakes are even more deadly than volcanoes. They often kill thousands of people when their houses crumble about them in a few seconds. Many people survive the earthquake itself but are buried alive and often badly injured. It is then a fight against time to rescue them before they die of suffocation or their injuries. Many cities in earthquake zones have well-trained rescue teams. But when disaster strikes in remote villages it can be days before any teams can reach them. Often the roads to the villages have become impassable. All earthquake rescue work is hazardous. Aftershocks can bring down damaged buildings on the rescuers. Fire may break out from fractured gas pipes, and there may not be enough water for firefighting because of burst water mains. There can also be great danger of disease from the decaying bodies of possibly thousands of people and animals.

Body heat
This is a picture taken by a thermal imaging camera. It records heat, not light. Rescuers use these cameras when searching in dark places for earthquake survivors.

With bare hands
A survivor of the 1995 Kobe earthquake in Japan uses his bare hands to remove debris. He is searching for other members of his family who might be buried in the ruins.

Ash and mud
An aerial view of Plymouth, the capital of the Caribbean island of Montserrat, after the volcanic eruption of 1997. Thick ash and torrents of mud have covered the city.

Still alive

Rescuers have heard faint cries from the rubble of a collapsed block of houses. Carefully, they remove the broken concrete and steel girders and find a survivor.

Stretcher bearers

Four members of the skilled rescue team that battled with the devastation caused by the Kobe earthquake in Japan, achieve another success as they carry a survivor to safety.

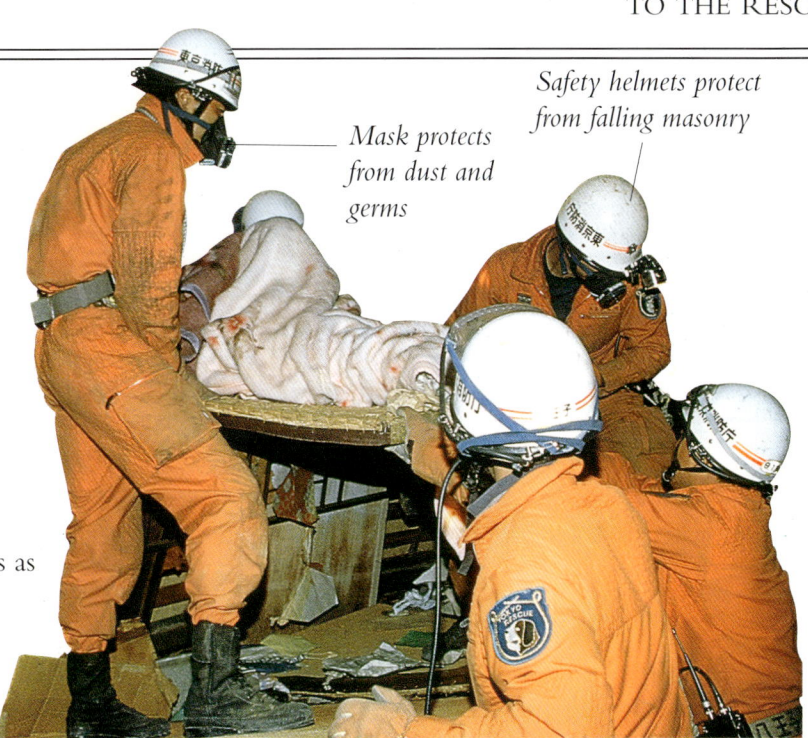

Mask protects from dust and germs

Safety helmets protect from falling masonry

Clearing up

Powerful excavators work round the clock to clear away shattered concrete and twisted metal supports from the collapsed Hanshin expressway after the Kobe earthquake in Japan. It took nearly a week to get the road back to normal.

Rescue man's best friend

Earthquake rescue teams not only rely on the latest scientific equipment to find survivors but also use sniffer dogs. Sniffer dogs are specially trained to use their sensitive noses to pick up the scent of people buried in collapsed buildings after an earthquake.

Heavy lifting

A crane is lowered to help lift the heavy steel and concrete beams of a collapsed building in Erzican, Turkey. These rescue workers are trying to free people buried during an earthquake there.

WILD WEATHER

*There are many kinds of weather – rain, snow,
sunshine, storm, frost and much more.
All of them are simply changes in the air
as it is stirred by the warmth of the Sun.
Look on a satellite picture of the world
and you can see some of these swirls and eddies
in the air picked out by the clouds.
It all seems rather chaotic, but
meteorologists, the scientists who study weather,
are at last beginning to understand
the atmosphere and just what makes
the world's weather machine tick.*

Author: Robin Kerrod
Consultant: Helen Young

WHAT IS WEATHER?

Some of the greatest challenges humans face are environmental disasters caused by the weather – from droughts and famines to blizzards and flash floods. Dealing with these conditions is an inevitable part of life on our planet. Sometimes the weather only affects our lives in a small way, such as in choosing what clothes to wear, or where to go on holiday. At other times, its consequences can be far more serious, as those who have seen the power of a tornado can testify.

Since the weather influences our lives in so many ways, scientists called meteorologists study patterns in the weather and try to forecast, or predict, what it is going to be like in the future. As research and technology advance, these predictions have become increasingly accurate.

Sense in the Sun

Many people find long, hot summer days a pleasant time of year, but the Sun can often burn your skin if you do not protect it. You should use suncream but, best of all, keep your skin covered up. Wear a sun hat to protect your head and sunglasses to cover your eyes.

Wrap up warm

Waterproof clothes, an umbrella and boots will protect you from the bad weather in most countries. But some places, such as the Arctic, are so cold that exposure to the icy temperatures there may be life-threatening.

Snow fun

Many people enjoy playing in the snow during the winter. But a heavy snowfall accompanied by a strong wind causes blizzard conditions, which make the outdoors a very dangerous place to be.

Dry as dust
Death Valley in California, USA, is one of the hottest places in the world. Rain sometimes falls and collects into small pools. But the water soon evaporates, leaving the cracked ground seen here.

Water everywhere
Heavy rain has caused flooding in the town of Kaskakia in Illinois, USA. Local rivers have burst their banks, and many people's homes have been submerged.

Winding up
A satellite picture reveals a tropical storm developing in the middle of the Pacific Ocean. Clouds spiral around the centre of an area of low pressure, driven by winds that may reach speeds of up to 200km/h.

Terror twister
A tornado powers its way through a small town in Texas, USA. The rapidly swirling winds of a tornado, or twister, can devastate buildings, uproot trees and toss vehicles into the air, cutting a path of destruction.

HEAT FROM THE SUN

Sun worship

An ancient Egyptian relief carved from limestone shows the pharaoh Akhenaten. He is offereing sacred lotus flowers to the Sun god, Aten. The ancient Egyptian people worshipped the Sun god because they recognized that life depended on the Sun. They kept track of time by watching the Sun rise and fall in the sky, and they realized that their crops depended on the Sun to survive.

THE Sun is a gigantic star that pours out vast amounts of energy, called electromagnetic radiation, into space. Although just a tiny amount of this energy reaches the Earth, it is enough to make rocks so hot that eggs can be fried on them. This heat energy also stirs the atmosphere into motion, powering the Earth's different weather systems.

Different parts of the Earth receive different amounts of heat from the Sun. The lower the Sun is in the sky, the less heat it provides. The Sun is directly overhead at the Equator, so it is much warmer here than at the Earth's poles. Some of the Sun's energy is reflected by the Earth's clouds, some by the ground and some by the atmosphere. The amount of heat that any one place receives from the Sun also changes from season to season.

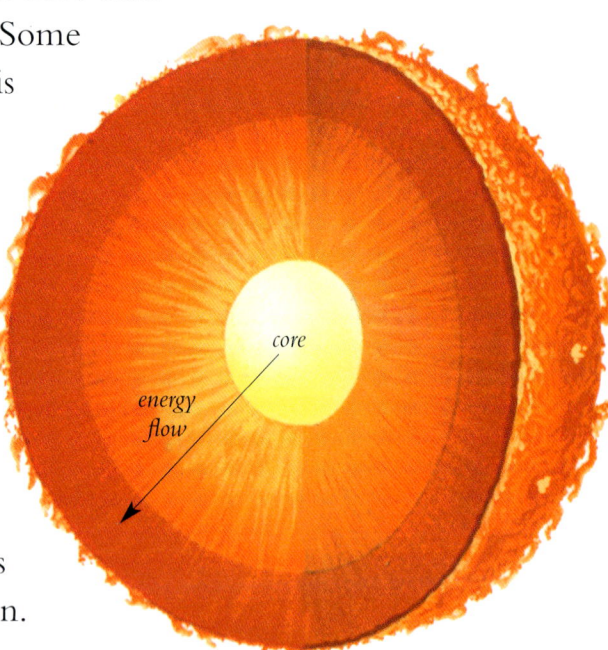

core

energy flow

Hot stuff

The Sun's energy is produced in its centre, called the core, where the temperature reaches 15,000,000°C. At this temperature, gases fuse (combine) to produce vast amounts of energy in the form of electromagnetic radiation, most of which is released as heat and light.

Heating the Earth

The Sun pours energy on to the Earth as heat and light. Some bounces off a thin blanket of gases, known as the atmosphere, back into space. The rest heats up the oceans, land and air. At night, clouds in the atmosphere help to stop heat escaping back into space.

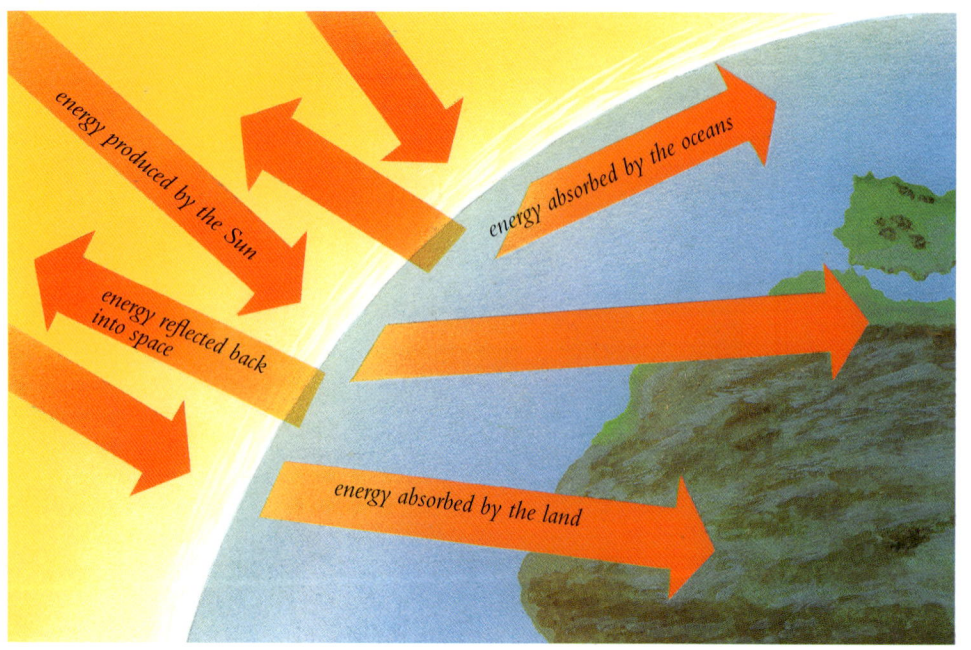

energy produced by the Sun

energy absorbed by the oceans

energy reflected back into space

energy absorbed by the land

Snowy December

During the winter in New York City, USA, it often snows in December and it is bitterly cold. The temperature may be just above freezing point (0°C), and a biting cold wind will make it feel even colder. New York is in the Northern Hemisphere. In this part of the world, the colder winter months begin around December. This is the time of year when the Northern Hemisphere is tilted most away from the Sun.

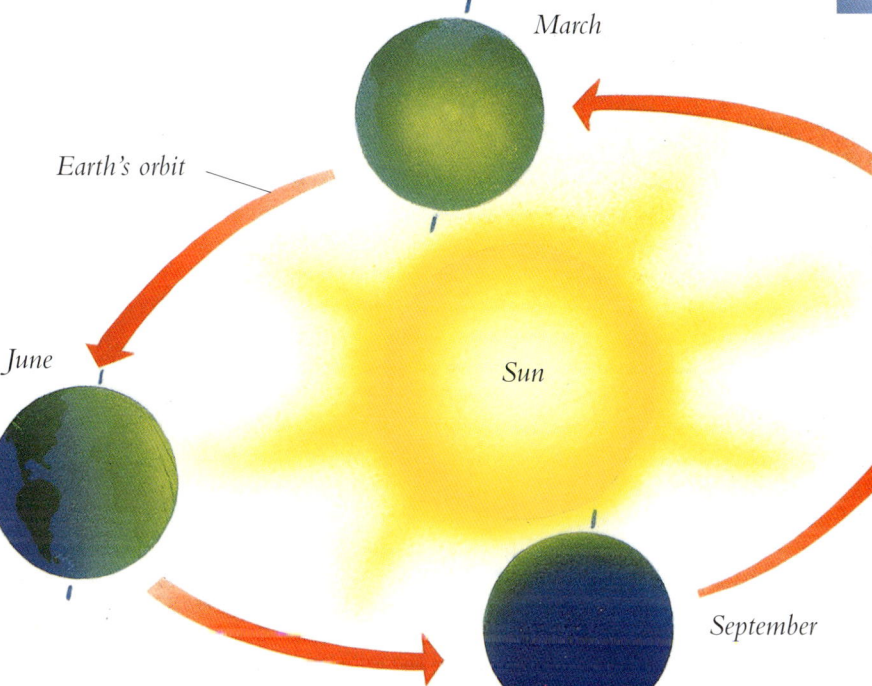

March

December

Earth's orbit

June

Sun

September

The changing seasons

Any one place on the Earth does not always receive the same amount of energy from the Sun. In fact, it changes with the seasons. Most energy is received in the summer season and least in the winter season. The seasons change because the Earth's axis is tilted in space. In summer, one half, or hemisphere, tilts towards the Sun and is warmer. In winter, the hemisphere tilts away from the Sun, and it is colder here.

Harnessing the Sun

A huge solar power plant in California, USA, converts energy from the Sun into electrical energy. Giant mirrors reflect sunlight on to a tank of water at the top of the tower. The water gets hot and boils, creating steam which drives giant turbines that generate the electricity. Solar power plants are useful in warm climates where the Sun shines steadily for most of the time.

Sunny December

In December, temperatures can exceed 30°C on beaches in Australia. Australia is in the Southern Hemisphere, where seasons are opposite to those in the Northern Hemisphere.

MEASURING TEMPERATURE

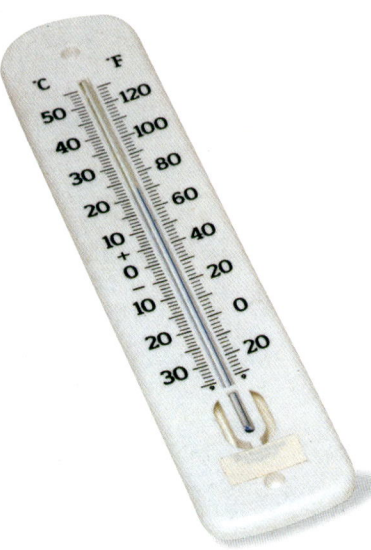

WHAT we notice most about the weather is the temperature — that is, how hot or cold it is. Temperature is measured using a thermometer. There are many different types of thermometer, but the most common consists of a glass column partially filled with a liquid such as alcohol or mercury (a liquid metal). When the temperature increases, the liquid expands in proportion to the rise in temperature and increases in length up the glass column. Similarly, a decrease in temperature causes the liquid in the glass column to decrease in length. This means that thermometers can record a range of temperatures. The simple thermometer in the experiment is made using water and can be used to record changes in temperature.

A simple thermometer
The temperature of the air can be measured using a simple mercury thermometer. Most have two temperature scales: degrees Celsius (°C) and degrees Fahrenheit (°F).

FACT BOX

• The human body is normally a constant temperature of 37°C.

• In 1922, the air temperature in Al´Aziziyah, Libya, rose to a sweltering 58°C in the shade.

• Water freezes into solid ice when the temperature falls below 0°C.

• Some parts of the Northern Hemisphere experience winter temperatures that regularly fall below −50°C.

• At a temperature of −190°C, all the air present in our atmosphere would turn into a liquid.

• When low temperatures combine with strong winds, our surroundings feel a lot colder than the temperature alone would suggest. This effect is known as the "wind-chill factor". For example, a combination of a wind speed of 48km/h and a temperature of just 4°C produces a wind-chill factor of −11°C, although water will not freeze in these conditions.

Taking your temperature
A thin strip of heat-sensitive material, called a thermo-strip, can be used to record the temperature of the human body. If you press the thermo-strip against your forehead, the heat of your body makes the strip change colour. The body temperature of a healthy human should be about 37°C.

Highs and lows
A maximum-and-minimum thermometer is used to check a range of temperatures inside a greenhouse. This type of thermometer indicates the highest and lowest temperature in 24 hours. If the temperature drops below 0°C, the girl's tomato plant will die.

MAKE YOUR OWN THERMOMETER

You will need: *cold water, plastic bottle, food colouring, straw, reusable adhesive, piece of card, scissors, felt-tip pen.*

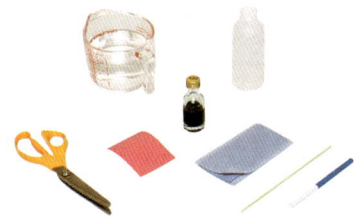

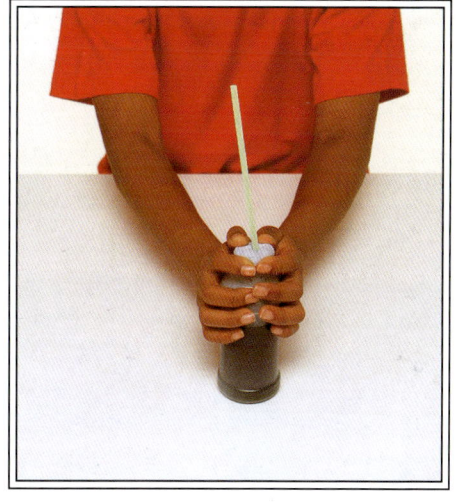

1 Pour cold water into the empty bottle until it is about two-thirds full. Add some food colouring. Dip the straw into the water and seal the neck tightly with reusable adhesive.

2 Blow into the straw to force extra air into the bottle. After a few seconds, the extra air pressure inside the bottle will force the water to move up inside the straw.

4 Then take your thermometer outside and leave it to stand for a while. On a hot day, the heat from the Sun will gradually make the air and water in the bottle expand. This will force water up the straw and past the level you marked for the room temperature. Mark the card again to show the temperature outside. Now put your thermometer in the refrigerator for two hours. The water level in the tube will drop below the room temperature mark. Make a note on your thermometer.

3 Cut two slots at either side of the card. Slide it over the straw. Leave the bottle to stand. Mark the card next to the water level to record the temperature of the room.

Stevenson screen

A meteorologist notes the temperature recorded by a pair of thermometers housed inside a shelter called a Stevenson screen. This protects the instruments from the weather. It is painted white to reflect sunlight and has louvred (slatted) sides to keep the air inside the shelter at the same temperature as the air outside.

CLIMATES OF THE WORLD

DIFFERENT parts of the world receive different amounts of heat from the Sun. As a result, they have a different weather pattern throughout the year. This changing pattern is called the climate. The world can be divided into regions with similar climates that suit different kinds of animals and plants.

In some regions near the Equator, the climate is hot all year round and plenty of rain also falls. Vast areas of tree-covered land, called rainforests, flourish there because of the heavy rainfall. On either side of the Equator, hot grasslands, called savannas, experience rain for only part of the year. Hot deserts are also found in this part of the world. Hardly any rain falls in deserts. Farther north and south, the climate is neither too hot nor too cold, and rain regularly falls. This is a warm temperate climate, common to most parts of the USA and Europe. However, the northernmost parts of North America, Europe and Asia have a cold temperate climate. The winters are long and cold, and plenty of snow falls. Evergreen forests dominate the landscape. In the far north of North America, Europe and Asia, it is too cold for trees to grow. These regions are called the tundra, and the temperatures may fall to −60°C. At the other end of the Earth, the continent of Antarctica has an equally cold climate.

Key to the climate
A world map can be divided into zones corresponding to the different types of weather patterns throughout the year. These are called climatic zones. Scientists classify climates in many different ways. The climate map below is divided into six different kinds of climate. Regions with different climates are inhabited by different types of animals and plants, which are well adapted to survive in their particular environments.

KEY

- tundra
- mountain
- cold temperate
- warm temperate
- desert
- tropical

Grazing the tundra
Caribou graze the thin vegetation of the Arctic tundra. The tundra has a harsh climate. In the winter it is extremely cold, and it only moderately improves in the summer. Most parts of the tundra are snowswept and frozen for up to nine months of the year.

In the tropics
Regions near the Equator have a tropical climate, which means it is very wet and warm. These conditions are ideal for rapid plant growth, creating rainforests. The trees and shrubs are evergreens, which means they keep their leaves all year long.

Arctic

tundra

conifer forest

deciduous forest

tropical forest

savanna

Mountain zones

Generally, the climate of a place is decided by its position on the Earth, but its altitude, or height above sea level, is also important. The temperature falls as you climb above sea level. Mountains in most parts of the world are layered with many zones, which have particular types of vegetation. These correspond with the climatic zones around the world.

Different types of mountains have different zonation. Generally, the Arctic zone is at the top, followed by tundra, then coniferous forest (evergreen trees with needle-shaped leaves), then deciduous forest (where the trees lose their leaves in winter). As the climate warms at lower levels, the vegetation becomes tropical forest, and finally savanna at the bottom.

An imaginary line called the snow line divides the Arctic and tundra zones. Above this line, there is a year-round cover of snow. Another imaginary line, called the tree line, divides the tundra and coniferous forest zones. Above this line, trees do not grow.

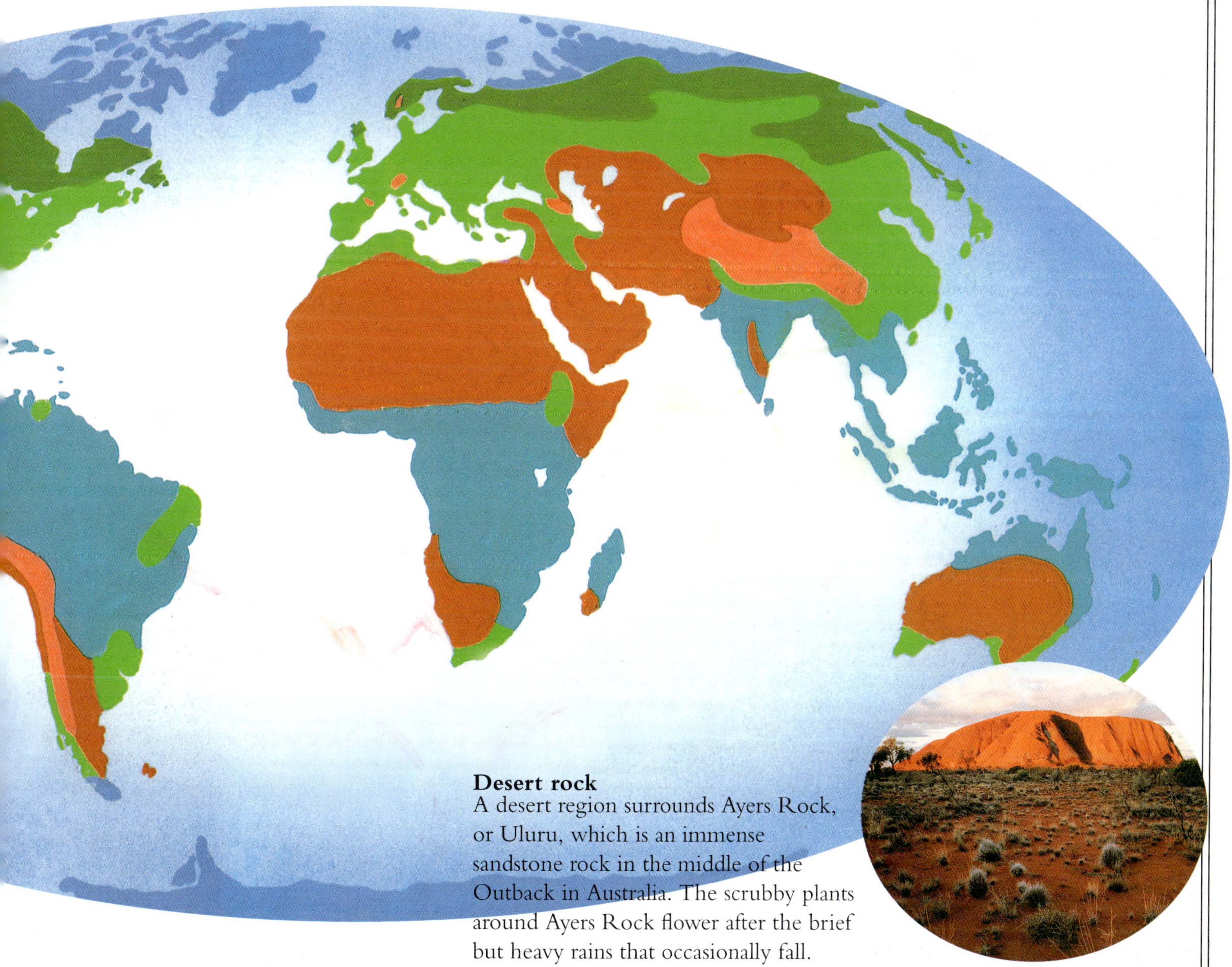

Desert rock

A desert region surrounds Ayers Rock, or Uluru, which is an immense sandstone rock in the middle of the Outback in Australia. The scrubby plants around Ayers Rock flower after the brief but heavy rains that occasionally fall.

CHANGING THE TEMPERATURE

Cool colour

In hot places, such as this town in Spain, all the houses are whitewashed to reflect the sunlight and keep people cool. There is not a dark house to be seen, because dark houses warm up faster by day and cool down faster at night.

DARK AND LIGHT

You will need: two identical glass jars with lids (paint the outside of one black and the other white), sand, watch, thermometer, notebook, pen.

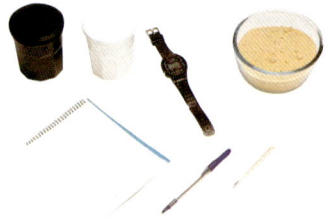

T HE TEMPERATURE of a place is controlled by different factors. The main factor is the amount of energy a place absorbs from the Sun, but other conditions play a part in controlling temperature, too. For instance, areas at very high altitudes are much colder than areas at sea level. Another factor is distance of a place from the sea. This is because the continents and the oceans do not heat up and cool down in the same way as each other. Water takes longer to heat up than the land, but the water holds its heat for much longer. Therefore, summers are cooler and winters are milder on the coast than they are inland.

Since water can circulate, it can move the heat from place to place in the form of ocean currents. Ocean currents therefore often affect air temperatures. For example, Britain and Labrador in Canada are the same distance from the Equator (they are the same latitude) but have very different climates. This is because a vast water current, called the Gulf Stream, transports warm water from the distant Gulf of Mexico to western Europe. This helps to keep winter temperatures much warmer in Britain than they are in Labrador.

1 Fill the two painted jars with sand to about the same level. Screw the lids on firmly. Place both jars outside in the sunlight and leave them there for about two hours.

2 Now take the temperature of the sand in each jar. The sand in the black jar will be hotter than the sand in the white jar. Write down the temperatures in your notebook.

3 Now put the jars in the shade. Note the temperature of the sand in each jar every 15 minutes. The sand in the black jar will cool down faster than the sand in the white jar.

PROJECT

MEASURING TEMPERATURE CHANGES

You will need: two bowls, jug of water, sand, watch, thermometer, notebook, pen.

1 Pour water into one bowl and sand into the other bowl. You do not need to measure the exact quantities of sand and water – just use roughly equal amounts.

2 Place the bowls side by side in a cool place. Leave them for a few hours. Then note the temperature of the sand and water. The temperature of each should be about the same.

5 In this experiment, the sand acts like land and the water acts like ocean. The sand gets hot quicker than the water, but the water holds its heat longer than the sand does.

3 Then place the bowls side by side in the sunlight. Leave the bowls for an hour or two. Then measure the temperatures of the sand and water in each bowl.

4 Then put each bowl in a cool place indoors. Record the temperature of the sand and water every 15 minutes. The sand cools down faster than the water.

Soaking up the Sun
Reptiles such as crocodiles are cold-blooded. They rely on the surrounding temperature to keep warm. Their dark-coloured skin helps them to absorb heat.

EL NIÑO AND LA NIÑA

GREAT masses of water are constantly on the move in all the world's oceans. They affect the climate of places along their path. For example, trade winds blow across the Pacific Ocean. Usually they are strong and blow from east to west. They drive a current of warm water westwards.

Every few years the trade winds suddenly weaken and the whole system goes into reverse. A warm ocean current, called El Niño (Spanish for boy child), appears along the coast of Peru. This reversal in direction of the ocean current has an alarming effect on the weather, causing heavy rain and flooding in some areas but drought and forest fires in others. After a while, the trade winds usually regain their former strength and things return to normal.

Sometimes, however, the trade winds become much stronger than usual and drive the warm Pacific Ocean current much farther west than usual. This reverses weather patterns, creating droughts in normally wet regions and floods in normally dry ones. This strong reverse current is known as La Niña (girl child). Both El Niños and La Niñas seem to be happening more regularly. Scientists are still trying to work out why this is so, but it could be caused by a gradual increase in the world's temperature, known as global warming.

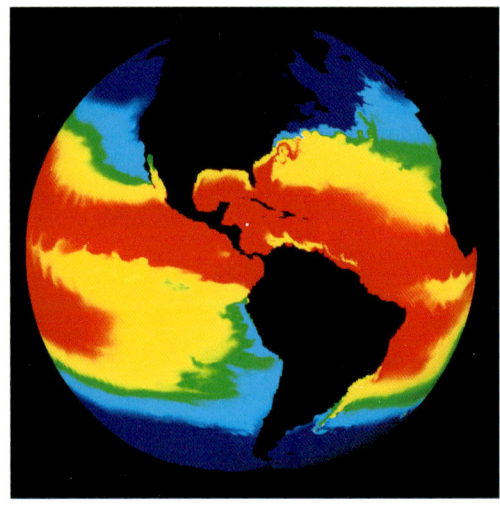

Ocean currents
A false-colour map shows how the temperature varies across the Earth's oceans. The warmest areas are red, followed by yellow, green and light blue. The dark blue areas near the North and South poles are the coolest areas. North and South America are the distinctive black shapes in the middle of the picture. The temperature of the oceans is due to the absorption of heat energy from the Sun. As the waters heat up, they begin to move and form oceanic currents. These currents are partly responsible for the Earth's weather patterns.

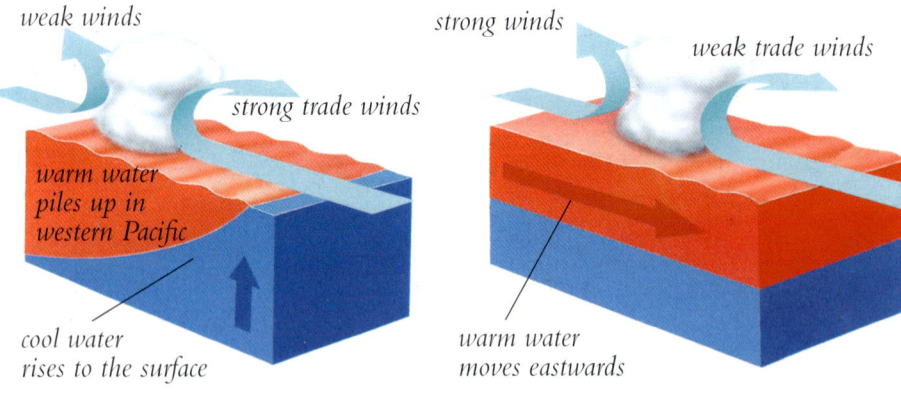

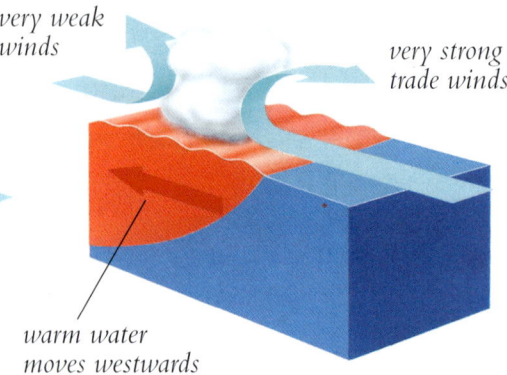

Normal conditions
In normal years, the steady trade winds of the tropical regions near the Equator blow westwards across the Pacific Ocean. They pick up moisture from the warm oceans and deliver rain to countries along the western Pacific Ocean.

El Niño
During El Niño years, the trade winds blowing westwards across the Pacific Ocean become much weaker than usual. The warm surface waters of the Pacific Ocean are forced eastwards by strong winds, which carry stormy weather systems.

La Niña
During La Niña years, the trade winds blowing westwards become very strong indeed. The warm surface water of the Pacific Ocean is forced westwards, leading to storms and much heavier rainfall than usual around the western Pacific Ocean.

Stormy beach

Houses just inland from Huntingdon Beach near Los Angeles, California, in the USA, have been flooded by heavy rain. Precipitation (rainfall) levels in California are very much affected by El Niño years. The Californian coast is normally dry and sunny. El Niño may cause a vast increase in precipitation but may also give rise to drought here. The El Niño of 1982–3 was particularly destructive. The cost of the damage was estimated to be up to $13 billion.

Fire hazard

In 1998, forest fires raged out of control in Indonesia, in South-east Asia. Usually, tropical rains blanket the islands of South-east Asia, but El Niño was responsible for a huge drought and subsequent forest fires. More than 20,000sq km of forests were destroyed, creating clouds of smoke and haze that spread to neighbouring countries such as Malaysia.

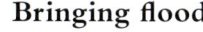

Bringing flood

El Niño has caused widespread flooding in Khartoum, the capital of Sudan in eastern Africa. The climate of Sudan is hot with seasonal rains during normal years. In El Niño years, however, the seasonal rains can be exceptionally heavy, flooding huge areas of land and making tens of thousands of people homeless.

Bringing famine

In the El Niño of 1997 and 1998, flooding devastated the crops in Sudan. Millions of people faced starvation and were forced to leave their homes. As a result of the flooding, relief agencies supplied food aid in camps such as this one.

BEATING THE HEAT

WEATHER and climate affect human life in many ways. More people live in warm climates than in colder climates. In some places, long dry spells, especially when accompanied by high temperatures, can lead to a shortage of food and even widespread starvation. In fact, parts of Africa have been in the grip of major drought and famine for decades.

Humans have developed a number of ways to avoid the effects of hot climates. Loose, light clothes help air to circulate around the body and protect the skin from sunburn, which can cause skin cancer. In the Mediterranean and parts of Asia, the walls of houses are very thick, keeping the house cool in the day and warm at night. Windows are small and covered with blinds or shutters to keep out the Sun. The rooms are tall so that the hot air rises to the ceiling.

Animals and plants from hot regions have adapted over time to cope with their surroundings and survive the heat.

Water shortage
When the weather is hot, the body needs plenty of water to avoid dehydration. When the body becomes too hot, it produces moisture in the form of sweat, which then evaporates into the air. This cools the skin and the blood beneath it.

The goat herders
Yemeni women in traditional dress lead a herd of goats to drink from a water supply in the Arabian Desert. Their hats and clothing cover their bodies and faces and protect their skin from the Sun. If they become too hot, the rapid loss of water through sweating may lead to heat exhaustion. The symptoms of this include sickness, tiredness and fainting.

Saharan landscape
A Bedouin tribesman gazes out over the Sahara Desert in northern Africa, a region in which the temperature regularly exceeds 40°C. He wears loose, light-coloured clothing, which covers him from head to foot. The man's clothes allow the air to circulate around his body and also reflect the sunlight. This protects him from the Sun's relentless heat.

Keeping cool

Running in and out of the spray from a garden sprinkler is a fun way to keep cool on a hot summer's day. Droplets of water from the sprinkler collect on the body and take heat away from the body as they evaporate. This makes the temperature feel lower than it actually is.

Desert survival

The camel is well adapted to life in its desert home. Its most famous feature is its fatty hump, which acts as a store of energy. This enables it to survive for many days without food and up to ten months without water. The camel produces very little urine, which cuts down water loss. The animal's thick fur also keeps it warm during the cold desert nights.

No sweat

The elephant's natural habitats are Africa and Asia. Its wrinkled and hairless skin retains water to help the animal cool down. The elephant does not sweat, but it can flap its two large ears to lose heat. The blood vessels in the ears are close to the surface of the skin and easily conduct heat away from the animal's body.

Prickly survivors

The huge trunk of the saguaro cactus is pleated like an accordion. After the rainy season, the stem absorbs water and the pleats unfold. A 6m-tall cactus can store up to a tonne of water in this way.

WHERE THE WEATHER IS

A lot of hot air
Hot-air balloons float quietly above the Masai Mara National Reserve in East Africa. A hot-air balloon is powered by a gas burner, which heats up the air inside the balloon. Warm air rises, so warming the air inside the balloon makes that rise too.

THE atmosphere is a layer of air that surrounds the Earth. Most of the air molecules are near the surface of the Earth. There are less air molecules the higher up you go, so the air is thinner there. At about 300km high, there are hardly any air molecules left in the atmosphere at all.

The Earth's weather mostly takes place in a layer of the atmosphere called the troposphere. This layer is between 10 and 16km thick. It is in this layer that clouds form, rain and snow fall and thunder and lightning take place.

In the next layer up of the atmosphere, known as the stratosphere, there is a layer of a gas called ozone. This blocks harmful radiation from the Sun. Recently, the ozone layer has been heavily damaged by harmful chemicals called pollutants, which include chlorofluorocarbons (CFCs). Despite a concerted worldwide effort to reduce their use, there is still deep concern that the ozone layer is thinning extremely rapidly. This thinning is especially noticeable over the North and South poles in spring.

Lights in the sky
Strange lights appear in the skies over the far north and far south, caused by particles from the Sun colliding with gases in the atmosphere. In the north they are called the northern lights or the aurora borealis. In the south they are called the southern lights or the aurora australis.

Into outer space
The region where the molecules of the Earth's atmosphere shoot off into space is sometimes referred to as the exosphere. This represents the upper limit of the Earth's atmosphere and occurs at around 450–500km above the surface of the planet.

Red sky

In the evening, the sky often turns red or orange. This happens because when the Sun is low in the sky, dust in the lower atmosphere scatters the blue light that we normally see. Only orange and red rays are left for us to see.

Blue sky

Skies appear blue because of the way light from the Sun is scattered by the molecules of gas in the air. Dust, water droplets and other particles reduce the intensity of the colours. The bluest skies are seen when the air is at its purest, away from pollution in the cities. Many people who live in cities travel out to the countryside to take part in sports and leisure pursuits, such as windsurfing. Cleaner air can help people to feel more refreshed.

Mountain high

The peaks of mountains often rise high above the clouds. At the top of a mountain, there are far fewer molecules of air in the atmosphere than at the bottom. The air is thin and lacks oxygen. This affects the functions of the human body, causing a condition known as altitude sickness. This may affect climbers and walkers at heights of around 3,500m and above. It causes feelings of sickness and light-headedness and, in severe cases, delusions or even death. This is the Aiguille Verte mountain in Chamonix, France, the peak of which is about 3,000m.

IN THE AIR

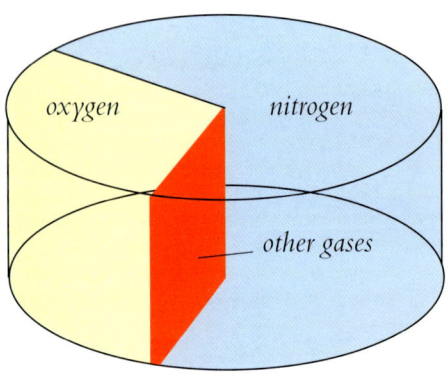

Gases in the air

About 78 per cent of the air is nitrogen and 21 per cent is oxygen The remaining 1 per cent is gases such as carbon dioxide and argon.

MEASURING THE OXYGEN

You will need: *candle, clear mixing bowl, reusable adhesive, jug filled with coloured water, glass jar, felt-tip pen.*

THE air we breathe is made up of a mixture of different gases. Nitrogen makes up most of the air's volume, but oxygen is the most important gas, because most living things need a constant supply of it to stay alive. We can work out the proportion of oxygen in the air in the simple experiment below. When things burn, they react with oxygen in the air and the oxygen is used up. As shown in the experiment below, if you burn a candle in a jar, you can use water to replace the oxygen that is used up. By noting how much water rises up the jar, you can estimate how much oxygen was in the jar to start with. You should find that the water level rises by about one-fifth, meaning that oxygen makes up about 20 per cent of the air.

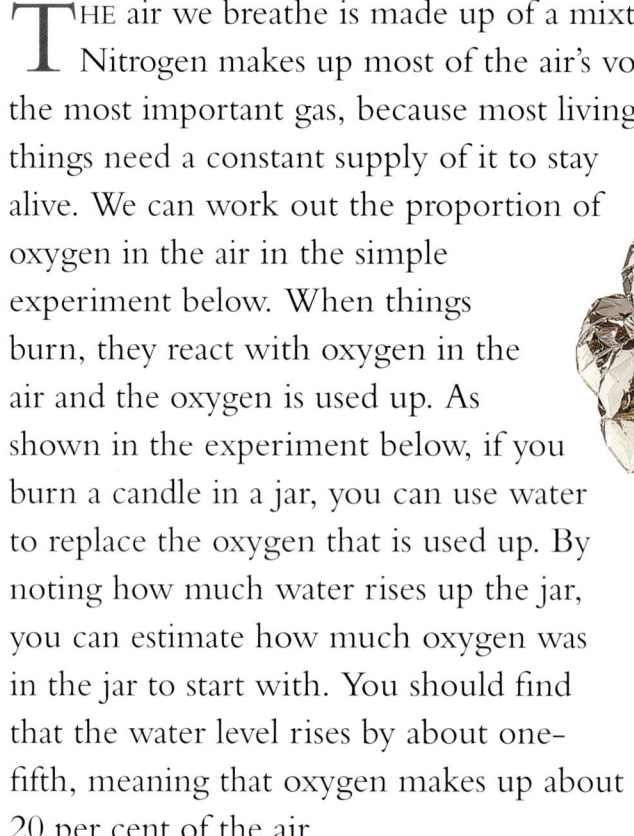

Up, up and away

A gas called helium can be used to inflate balloons. Helium-filled balloons float away quickly if you let go of their strings, because helium is lighter than the air.

1 Secure the candle to the bottom of the mixing bowl with reusable adhesive. Pour enough coloured water in the bowl to fill it to a depth of about 2–3cm.

2 Ask an adult to light the candle. As soon as it starts to burn, place the jar over the candle. Let the jar rest in the water on the bottom of the bowl and watch what happens.

3 The water rises from the bowl up into the jar until the candle goes out. Mark the water level on the jar – this will show how much oxygen was in the jar to start with.

P R O J E C T

SEE THE WEIGHT

You will need: *scissors, roll of sticky tape, ruler, piece of thread, two balloons of the same size, balloon pump.*

1 Using your scissors, carefully cut a small piece off the roll of sticky tape. Fix the tape around the middle of the ruler. Cut a piece of thread and tie it to the tape on the ruler.

2 Lift up the ruler by the thread and see if you can balance it horizontally. You will need to adjust the position of the thread until the ruler balances.

3 Take a balloon and blow it up a little. Take the other balloon and blow it up to a much larger size than the first. Carefully tape the balloons to opposite ends of the ruler.

4 Hold up the ruler with the thread. The large balloon makes the ruler dip down at one end. It is heavier than the small balloon, because it contains more air.

FACT BOX

• All gases weigh something, but some gases are heavier than others. For instance, hydrogen is heavier than helium.

• If our atmosphere consisted only of hydrogen gas, the balloon filled with helium would sink, rather than float away.

AIR ON THE MOVE

THE air in the atmosphere has weight. It pushes down on everything on the Earth with a force called atmospheric pressure. At sea level, atmospheric pressure is equivalent to the force of about 1kg on every square centimetre of the Earth's surface, but there are slight differences in atmospheric pressure from place to place. These differences make the air travel from a region of high pressure to a region of low pressure. This moving air is wind. Wind speed varies greatly, from a breeze (up to about 50km/h) up to a hurricane (up to 300km/h).

Breezes are gentle winds that often occur at the seaside where there is a difference in temperature between the sea and the land. The difference in temperature causes a difference in pressure and a breeze blows.

In contrast, large-scale wind systems blow all around the world, forming wind belts. These occur because of the differences in temperature between the hot Equator and the cold polar regions. The Earth is constantly rotating, and if this did not occur, the winds would blow from the high-pressure region over the cold poles to the low-pressure region near the Equator. Since the Earth rotates, however, this movement sets up a force called the Coriolis effect. This causes the wind to turn to the right of its path in the Northern Hemisphere and to the left of its path in the Southern Hemisphere. These two strong currents of air are called the trade winds, named because they once helped trading ships sail the oceans.

Sir Francis Beaufort
In 1805, English naval officer Sir Francis Beaufort (1774-1857) devised the scale that takes his name. The Beaufort scale soon became the standard method of estimating wind speeds, and it is still used today.

The Beaufort scale

The force, or strength, of the wind varies from place to place. The Beaufort scale is used to estimate the force of the wind. The scale is measured from Force 0, which means the air is not moving, to Force 12, which means a hurricane is blowing. One way you can guess the force of the wind is by the effect it has on you.

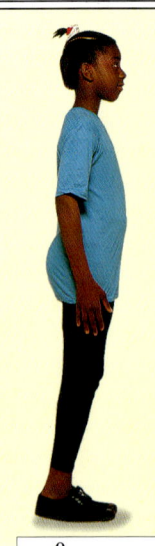

Force 0 on the Beaufort scale means that the wind speed is not noticeable. Smoke from a chimney rises vertically. When the wind reaches Force 2, you can feel the moving air on your face. This wind is called a breeze.

You can feel a Force 4 breeze pushing against your body when you walk.

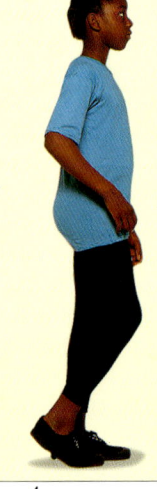

| 0 | 1 | 2 | 3 | 4 | 5 |

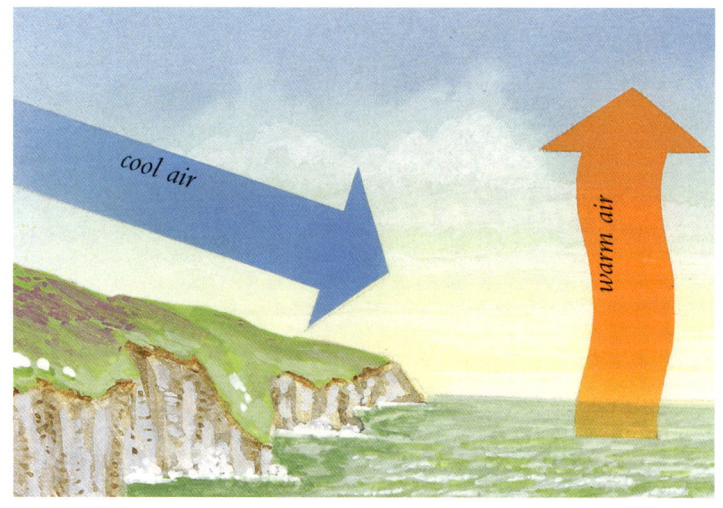

Land breezes and sea breezes

During the day, sea breezes are caused when the Sun heats up the land, and it becomes warmer than the sea. The warm air rises above the land, pulling in cool air from the sea. The opposite (shown above) happens during the night, when a land breeze blows off the land.

Go fly a kite

A young girl flies a kite at the seaside. A kite flies when air moves past it. The movement of the air produces an upwards force, called lift, on the kite. This force is responsible for supporting the weight of the kite and keeps it suspended in the air. The Chinese are thought to have invented the kite more than 2,300 years ago.

Wind power

Windmills were once used across Europe for milling (grinding) grain into flour. The enormous sails were powered by the wind. As the sails turned, they worked the machinery inside the mill.

Wind farm

Giant propellers harness the power of the wind at a wind farm in Altamont Pass in the USA. The propellers drive turbines that generate electricity for a nearby town.

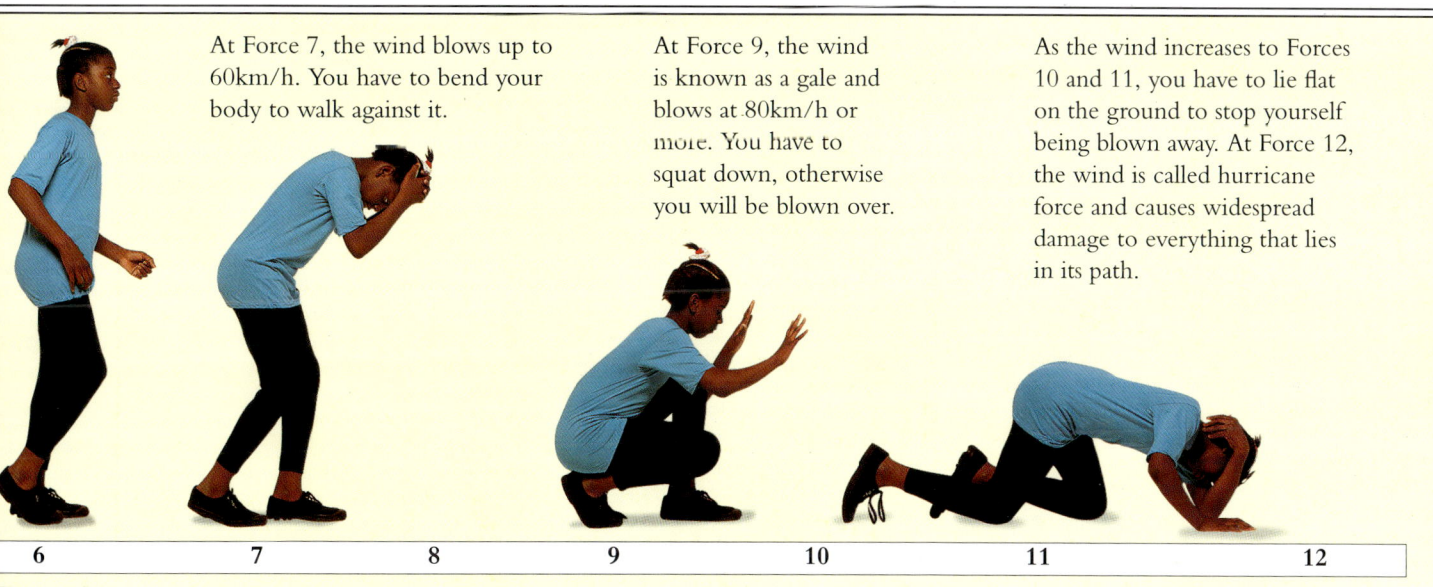

At Force 7, the wind blows up to 60km/h. You have to bend your body to walk against it.

At Force 9, the wind is known as a gale and blows at 80km/h or more. You have to squat down, otherwise you will be blown over.

As the wind increases to Forces 10 and 11, you have to lie flat on the ground to stop yourself being blown away. At Force 12, the wind is called hurricane force and causes widespread damage to everything that lies in its path.

6   7   8   9   10   11   12

AIR PRESSURE

THE weight of the air causes a force called air pressure to push down on the surface of the Earth. Air pressure causes air to move in the atmosphere, because the molecules in the air always move from areas of high pressure to areas of low pressure.

Air pressure varies according to many factors, such as air temperature and air density (how tightly its particles are packed together). The molecules in cold air move slower than the molecules in warm air and they crowd closer together. Dense cold air contains lots of molecules and puts a greater force on the Earth's surface.

Here are some tricks that involve air pressure. We can use the pressure of the air to knock a pile of books over. Sometimes paper can appear to be stronger than wood. This is due to the force of the air on the paper. In a fizzy chemical reaction, other gases, besides air, exert pressure on a balloon, forcing the balloon to expand.

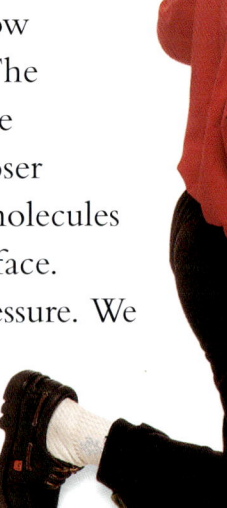

Pump it up
When you pump air into the tyres of a bicycle, you increase the number of air molecules in each tyre. The pressure of the air inside the tyre also increases – the tyre feels hard after you have pumped it up.

FEELING THE PRESSURE

You will need: balloon, books, balloon pump, wooden strip, newspaper, thick protective glove.

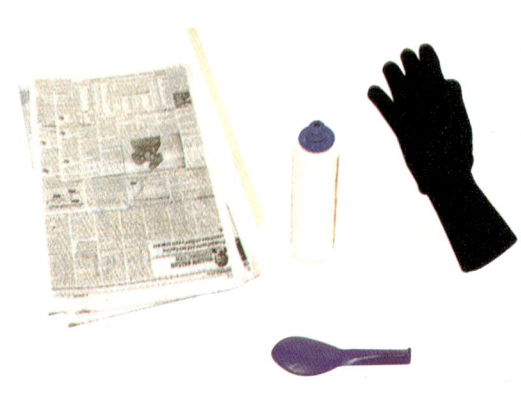

1 Place a balloon under some books. Blow air into the balloon using a balloon pump. As you pump, the air pressure inside the balloon rises. The increased force on the books pushes the pile over.

2 Cover the wooden strip with newspaper on a table. Leave a piece of the strip hanging over the end of the table. Wear a glove and strike the strip. Air pressure holds the paper in place and the wood snaps.

PROJECT

CREATE AIR PRESSURE

You will need: *funnel, bottle, vinegar, balloon, baking powder.*

1 Place the funnel in the neck of the bottle. Carefully pour in some of the vinegar, up to about half way. Make sure you wash the funnel after you have used it.

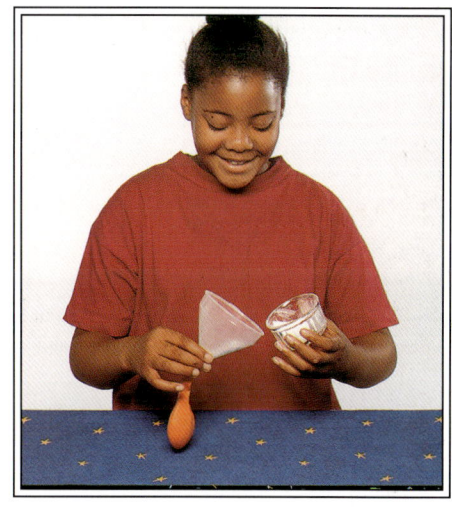

2 Fit the opening of the balloon around the bottom of the funnel. Carefully tip some of the baking powder into the funnel. Shake the powder into the balloon.

3 Carefully fit the opening of the balloon over the neck of the bottle of vinegar. Let the powder-filled part of the balloon hang down to one side of the bottle.

4 Gently turn the balloon over, so that the baking powder tips into the vinegar. The balloon expands as the mixture starts to fizz.

5 The fizzing indicates that a chemical reaction is taking place between the vinegar and the baking powder. This reaction produces lots of a gas called carbon dioxide. The pressure inside the balloon rises as more carbon dioxide molecules fill it. As a result, the increasing pressure of the carbon dioxide forces the balloon to expand.

WHIRLWINDS AND TORNADOES

Whirling wind
Dust devils spring up on dusty land in hot, dry summers. They pick up dirt and carry it high into the sky. The wind rotates relatively slowly in a dust devil, and so does little if any damage.

WHEN you see leaves spinning round and round in the breeze, you are witnessing a miniature whirlwind. A similar thing happens on dusty land in the summer. Little whirlwinds pick up the dry dust and spin it around and upwards into a spiralling column called a dust devil, which can climb up to 300m high.

Dust devils do little damage, but another kind of whirling wind is one of the most destructive forces in nature. It is the tornado, nicknamed twister because of its rapidly rotating winds. Tornadoes are born in violent thunderstorms. They form when a funnel-shaped column of whirling air forms beneath a thundercloud and descends to the ground. As it nears the ground, it picks up dust and debris. With winds racing round at speeds of up to 500km/h, a tornado destroys everything in its path, ripping houses to pieces and tossing cars into the air. A typical tornado is about 300m across and moves across the ground at a speed of about 50km/h.

Tornadoes occur regularly on the flat central plains of the USA, mostly along a broad path through Texas, Oklahoma, Kansas and Missouri. This area has become known as Tornado Alley. Tornadoes also form at sea, when they are called waterspouts. They are not as powerful and only last for only a few minutes.

A tornado is born
A huge thundercloud develops near Toronto, Canada. It is difficult to predict if a thunderstorm will give rise to a tornado, but meteorologists can tell where tornado-generating storms are most likely to form, and will issue a tornado warning.

Sinister silhouette
The setting sun highlights the dark shape of a tornado in Colorado, USA. The rapidly rotating column of air that touches the ground is known as the mesocyclone.

Fearful funnel
A dark, spinning funnel of a well-developed tornado heads for a town in Texas. If it hits the town, it will carve out a path of destruction several hundred metres across.

Furious freak
A freak tornado in Windsor Locks, Connecticut, USA, has left many houses in ruins. Tornadoes are most common in the Central Plains region of the USA. Here, severe thunderstorms often develop, and these are ideal for the development of tornadoes.

Watch the waterspout
Waterspouts are much like tornadoes, but they occur over water rather than the land. They are common in all equatorial oceans and inland seas. The water in the spout is formed by water vapour in the air condensing into water droplets. These are then pulled into the updraft within the cloud. Unlike tornadoes, however, waterspouts are usually rather weak storms and rarely cause much damage.

Tossed aside
Winds blowing at more than 300km/h have hurled an aircraft on to a nearby barn during a tornado in Louisiana, USA.

MEASURING THE WIND

THE wind shifts air from place to place and brings about changes in the weather. Meteorologists chart the direction and speed of the wind to help them predict these changes. They use an instrument called a weather vane to find out the wind direction. Weather vanes are often made in the shape of cockerels, when they are called weather cocks. The project on this page tells you how to make a simple weather vane.

To measure the wind speed, meteorologists use a device called an anemometer. Most consist of a circle of cups that spin round when the wind blows, rather like a windmill. The faster the wind blows, the faster the anemometer spins.

MAKE A WEATHER VANE

You will need: *reusable adhesive, plastic pot and its lid, scissors, garden stick, plastic straws, coloured card, pen, sticky tape, pin, plywood, compass.*

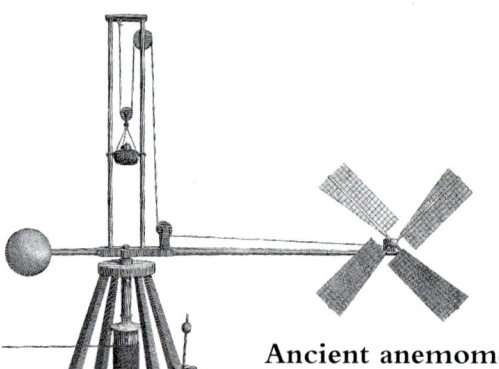

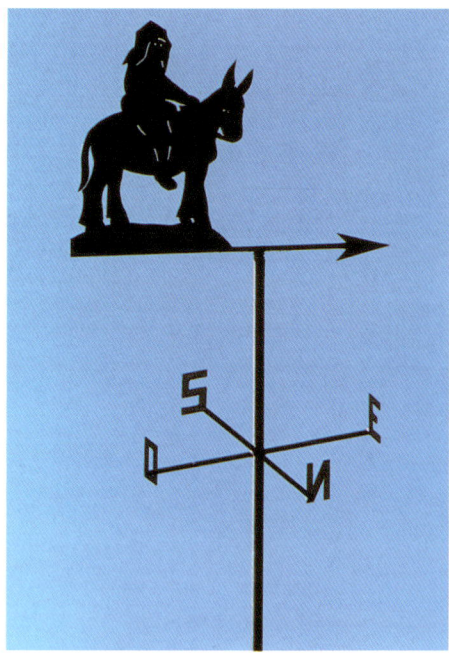

Going west
Weather vanes are commonly found on church steeples. This one points towards the east, which tells us that the wind is blowing from the east.

Ancient anemometer
The arm of this French anemometer from the 1600s moved when the wind blew against the propeller on the arm. The amount the arm moved was a measure of the wind speed.

1 Stick a ball of reusable adhesive to the middle of the lid of the pot. Ask an adult to pierce a hole in the bottom of the pot with the scissors. Place the pot on to the lid.

2 Slide the stick into one straw. Trim the end of the stick so that it is a little shorter than the straw. Push the straw and stick through the hole in the pot and into the adhesive.

3 Cut out a square of card. Mark each corner with a point of the compass – N, S, E, W. Snip a hole in the middle of the card and carefully slip the card over the straw.

PROJECT

4 Cut out two card triangles. Stick them to each end of the second straw to form an arrow head and tail. Put a ball of reusable adhesive in the top of the first straw in the pot.

5 Push a pin through the middle of the arrow. Stick the pin into the reusable adhesive in the first straw. Be careful not to prick your finger when you handle the pin.

6 Secure your weather vane to a plywood base using a piece of reusable adhesive. Test it for use – the arrow should spin round freely when you blow on it.

7 When you have finished your weather vane, take it outside. Use a compass to make sure the corners of your weather vane point in the right directions. You can then use the weather vane to find out the direction the wind is blowing.

Windmills

The miniature windmills on the toy above spin faster the harder you blow on them. The sails of real windmills also spin faster as the speed of the wind increases. As a result, windmills need a "governor" to regulate the speed of their rotation so that the sails are not damaged in strong winds.

Wind direction

Don't forget that the arrow points in the direction that the wind is blowing from. So if it points west, the wind is a west wind.

HURRICANES AND CYCLONES

THE whirling storms we call tornadoes are extremely powerful, but they are only a few hundred metres across and travel only a few kilometres. Much larger and more destructive are the great whirling storms called tropical cyclones. They form over the oceans of tropical regions. Gradually, they grow into great spirals of dense clouds as much as 500km across, with winds whirling round at speeds up to 300km/h.

Tropical cyclones that form north of the Equator and in the oceans around the USA are called hurricanes. They are born in the warm waters of the Atlantic and East Pacific oceans and often affect the USA and the Caribbean. Elsewhere in the world, hurricanes have other names. For example, typhoons form in the North Pacific Ocean and affect Japan.

When hurricanes hit land, they unleash heavy rain and howling winds that cause massive destruction. Strangely, in the centre of a hurricane there is a calm area about 30km across, called the eye, where there is little wind or cloud.

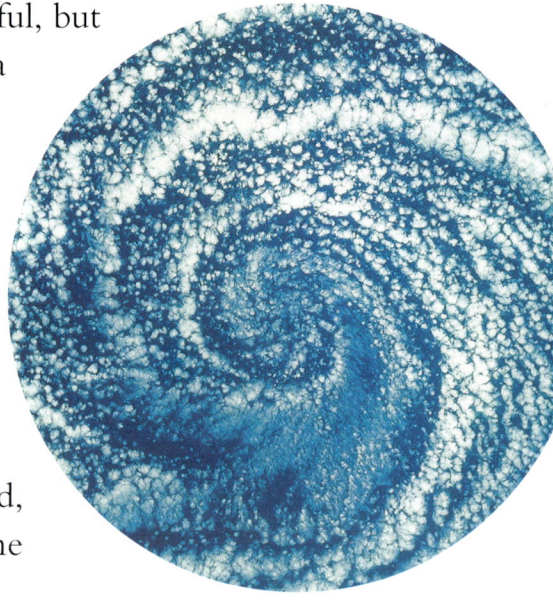

In a spiral
Space shuttle astronauts photographed a cyclone in the eastern part of the Pacific Ocean, off California, that covered hundreds of square kilometres.

Hurricane Gilbert
Although the high winds of a hurricane can inflict a great deal of damage, the majority of destruction is usually brought about by huge tidal waves and flooding. In September 1988, a total of 200 people died when Hurricane Gilbert slammed the Gulf Coast of Mexico.

One in the eye

An image taken by the crew of a space shuttle in November 1991 shows the clearly defined eye of a cyclone known as Typhoon Yuri. This typhoon formed near the Philippines in the eastern Pacific Ocean. Gradually, Typhoon Yuri grew to be more than 1,700km in diameter. The clouds lining the wall of the typhoon extended to between 13km and 15km deep. A typhoon of this huge size would give rise to winds with speeds of more than 250km/h with sudden blasts (gusts) of over 270km/h – a truly awesome spectacle of nature.

Mitch's mudslides

Although the high winds of a hurricane can inflict a great deal of damage, it is usually the huge waves and associated flooding that cause the most damage. Heavy rain following the passage of Hurricane Mitch caused huge mudslides in Tegucigalpa, Honduras, in late October 1998. In total, 17 people were killed as a result of the mudslides alone.

Andrew the destroyer

Some of the 200,000 or more homes and businesses in southern Florida, USA, that were destroyed or severely damaged by Hurricane Andrew in August 1992. The hurricane left 65 people dead, over 160,000 homeless and caused about $30 billion in damages.

Path of devastation

A time-lapse satellite image of Hurricane Andrew shows the path it took across the Atlantic Ocean and across southern Florida, USA. The passage of the hurricane from right (23 August 1992), to middle (24 August 1992) and left (25 August 1992) was monitored by meteorologists. Warnings were issued but, although important, they are not always able to save lives and property.

MASSES OF AIR

Great bodies of air, called air masses, are moving through the atmosphere all the time. These air masses can be huge, often covering whole continents or oceans. The way they move is complicated. It depends on the air pressure and the density of the air mass, and the Coriolis effect caused by the Earth spinning around in space.

Air masses are associated with particular types of weather, because they have certain temperatures and contain certain amounts of moisture. While a single air mass is passing us by, the weather remains the same. When another air mass takes its place with a different temperature and moisture, the weather changes. The worst weather occurs when a cold air mass and a warm air mass meet. Thick clouds form at this boundary, called a front. The weather settles down again after the front has passed by, that is, until another air mass replaces it.

Stormy weather
A storm may occur if two different air masses collide with one another. If a mass of warm air meets a mass of cold air, it produces widespread cloudiness and rain.

In the doldrums
Over the Equator, the air is warm and the winds are light. Many years ago, sailing ships were unable to move for days when they came upon these regions, called the doldrums. However, storms can suddenly spring up in the doldrums due to the upward surge of moist air heated by the ocean.

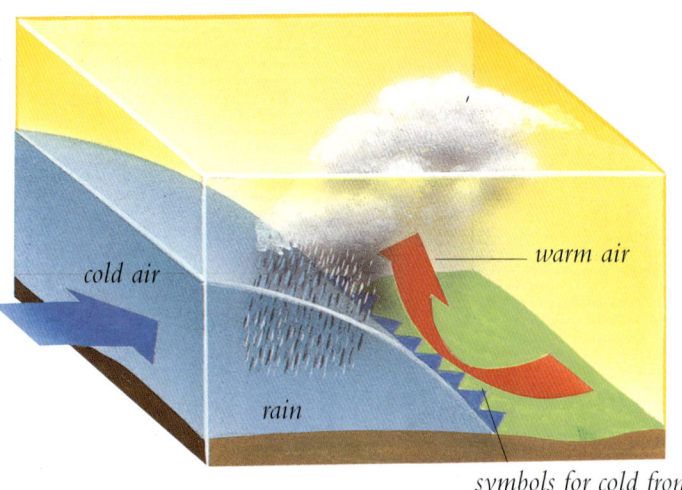

warm air

cold air

rain

symbols for cold front

Cold front
The meeting point between two air masses is called a cold front. It forms as a wedge of cool air pushes underneath a mass of warm air. The warm air is forced upwards. As it cools, clouds form and rain falls.

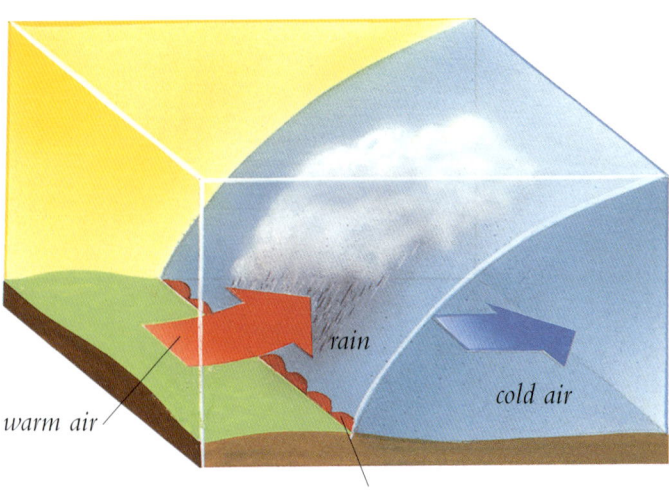

rain

cold air

warm air

symbols for warm front

Warm front
When a mass of warm air rides up over a mass of cold air, it causes a warm front. The rising of warm air over cold air, called over-running, produces clouds and rain well in advance of the front's surface boundary.

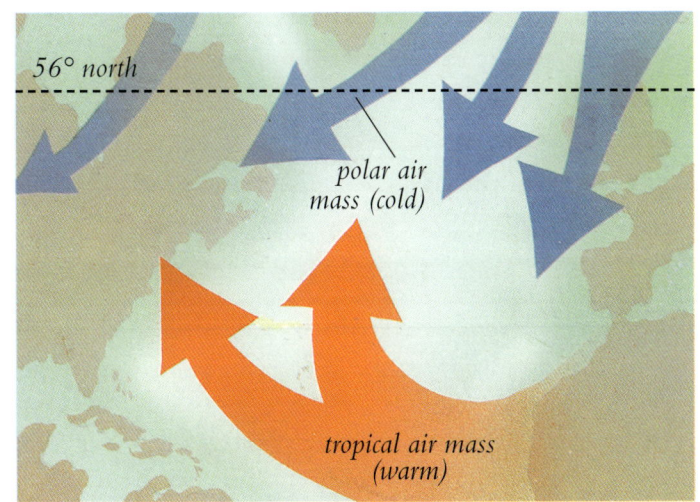

56° north

polar air mass (cold)

tropical air mass (warm)

Polar fronts

Great masses of warm and cool air move around the Earth. Warm air masses travel north and south away from the Equator. Cold air masses travel south and north away from the North and South poles. As a mass of warm air travels towards the poles, it encounters cold air moving down from the poles. The warm and cold air masses do not readily mix and they are separated by a boundary called the polar front. The polar front features low air pressure, called the sub-polar low, where surface air piles up and rises and storms develop. High in the sky, some of the rising air moves back towards the Equator, where it sinks back to the Earth's surface. This air then flows back towards the poles and the process is repeated.

KEY

westerlies

trade winds

polar easterlies

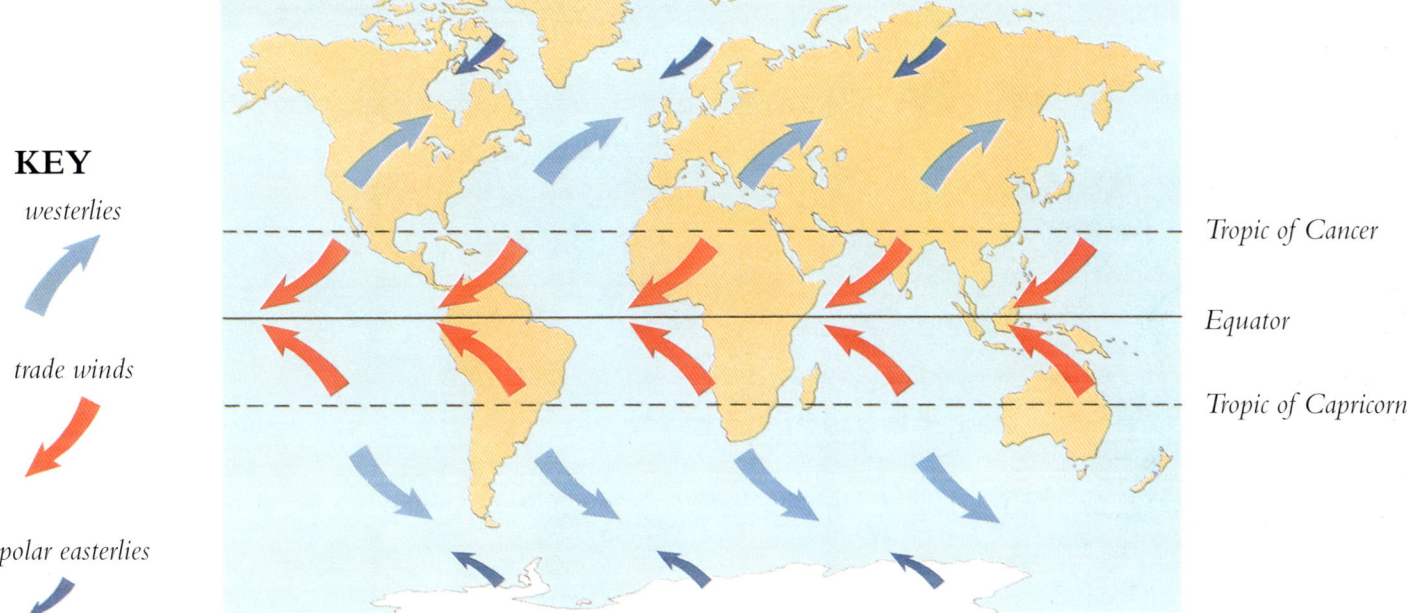

Tropic of Cancer

Equator

Tropic of Capricorn

World winds

This map shows the major movements of air around the world. They are called wind belts. The winds that blow in the belts are prevailing winds, which means they nearly always blow in the same direction. The trade winds are warm because they blow either side of the Equator. The westerlies are cool and blow north of the Tropic of Cancer and south of the Tropic of Capricorn. The polar easterlies are icy and blow round the North and South poles.

Fast winds

Sailing boats and ships rely on the trade winds and westerlies to cross the oceans. In fact, traders have relied on the trade winds for thousands of years to transport their ships across the Indian Ocean.

THE WATER CYCLE

Water moves around the Earth and its atmosphere in a continuous process called the water cycle. Heat from the Sun transforms water from oceans, lakes and rivers, into a gas, called water vapour, in a process called evaporation. In the atmosphere, water vapour rises, then cools and changes back into tiny droplets of liquid water. This is called condensation. Water droplets then gather together to form clouds. When water in the atmosphere is too heavy to be held in the air, it comes back to the Earth's surface as precipitation – dew, rain, sleet or snow.

On the surface, water can be consumed by animals or taken up by plants. Plants use as much water as they need and then release the rest back into the atmosphere in a process called transpiration. Sometimes water sinks below the Earth's surface to replenish underground supplies of water called groundwater. It can also remain on the surface in rivers and streams or lie frozen as glacial ice. Eventually, the water in lakes, rivers, streams and oceans evaporates once more to complete the water cycle.

Beneath the surface
When it rains, the water does not always flow into lakes, streams or oceans. Some disappears into the ground and becomes groundwater. In some places, this forms huge reservoirs of pure water hundreds of metres below the surface of the Earth. Underground caverns form in limestone rock by the erosion (wearing away) of the rock by groundwater.

The water cycle
The Greek thinker Thales of Miletus (c.625–c.550BC) was the first to describe the water cycle, over 2,500 years ago. The four main stages are evaporation, transpiration, condensation and precipitation. They form a continuous cycle.

precipitation

condensation

Sun

transpiration

evaporation

Elk in a fog

An elk looks lost in the early morning fog. When the visibility drops to less than 1km, the air is wet with tiny floating water droplets. This wet haze becomes a cloud resting near the surface of the Earth and is called fog. It is formed when warm air full of water vapour moves in above cold ground. The vapour then condenses into droplets of water.

City in a smog

A reddish brown smog smothers Hong Kong. Smog is caused mainly by the exhaust gases from the motor vehicles that clog the city's streets. These gases contain unburnt particles, which combine with a gas called ozone to form the hazy smog.

Mountain stream

A stream runs through a valley in the Rocky Mountains. On the horizon, puffy white clouds drift towards the high peaks. As the clouds pass over the mountain peak they develop into rain clouds. As the clouds pass down the other side of the mountain, they shed their water in the form of rain or snow. This will feed the stream with a constant supply of water.

Turning to vapour

Sunlight streams through a hole in a thick cloud off Maui Island in the Pacific Ocean. The Sun heats up water from the tropical ocean, and the water evaporates as a vapour in the air. As the vapour rises it cools and condenses, causing the cloud to grow.

HUMIDITY

WHY does 21°C in the Caribbean feel so much hotter than 21°C in Egypt? The answer is humidity – the air's water vapour content. When there is a lot of water vapour about, such as in the Caribbean, the air feels moist and sticky. This is because the perspiration on our skin cannot evaporate into the air very well – there is too much water in the air already. As a result, the perspiration stays on our skin and makes it wet, preventing us from cooling down. When there is little water vapour about, such as in Egypt, the air feels dry. The perspiration on our skin escapes into the air more easily.

Measuring the amount of water vapour in the air helps meteorologists forecast the weather. When the air is very humid, there is more chance that it will rain. Meteorologists use a device called a hygrometer to measure humidity. You could make a simple hygrometer using a long hair from your head. The length of hair changes as the humidity changes. This is the basis of an instrument called the hair hygrometer.

Another device meteorologists use to measure humidity is the wet-and-dry bulb thermometer, which contains two different thermometers. The difference in temperature between the two thermometers is used to calculate the humidity. You can make a simple hygrometer in this project.

Water producers
In the rainforests along the coast of northern California, USA, it is warm and humid most of the time. Rainforests contain thousands of plants, all of which give off vast amounts of water from their leaves.

MEASURE THE HUMIDITY

You will need: 2 sheets of coloured card, pen, scissors, glue, toothpick, used matchstick, straw, reusable adhesive, blotting paper, hole punch.

1 Cut out a card rectangle. Mark regular intervals along one side for a scale. Cut a 2cm slit in one short side. Split the parts out as shown above and glue them to a card base.

2 Cut another long rectangle from the first card. Fold it and stick it to the card base as shown above. Pierce the top carefully with a toothpick to form a pivot.

P R O J E C T

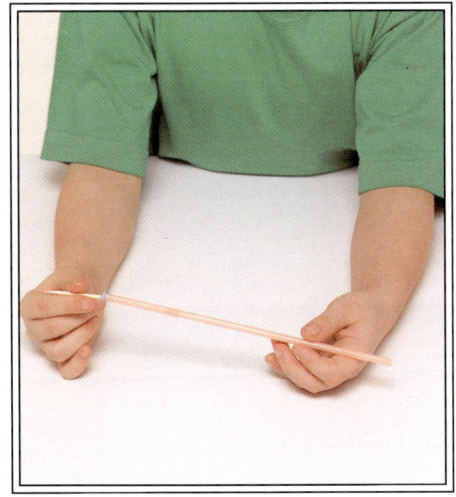

3 Fix the used matchstick to one end of the straw using some reusable adhesive to make a pointer. Both the matchstick and the adhesive give the pointer some weight.

4 Carefully cut out several squares of blotting paper. Use the hole punch to make a hole in the middle of each square. Slide the squares over the end of the pointer.

5 Now carefully pierce the toothpick pointer with the pivot. Position the pointer as shown above. Make sure the pointer can swing freely up and down.

6 Adjust the position of the toothpick so that it stays level. Take the hygrometer into the bathroom when you have a bath. The high humidity should make the blotting paper damp and the pointer will tip upwards. On a warm day outside, the blotting paper will dry and the pointer will tip down.

Transpiring plants

Plants play a vital role in the transfer of water from the atmosphere to the Earth. A plant's leaves give off water vapour in a process called transpiration. Cover a pot plant with a clear plastic bag, sealing the plastic around the pot with sticky tape. Put the plant in direct sunlight for two hours. Notice that the bag starts to mist up and droplets of water form on the inside. They form when the water vapour given off by the plant turns back to a liquid.

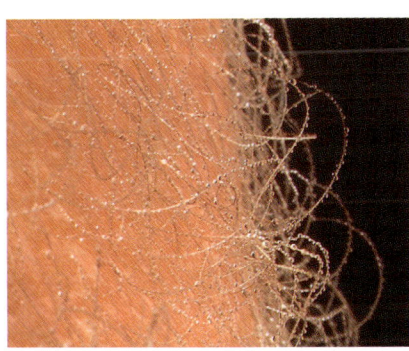

Building a sweat

Perspiration is the process by which we cool down. Heat is removed from the skin when water (sweat) evaporates from it. This process is less effective in humid weather when the sweat remains on our skin.

LOOKING AT CLOUDS

Without clouds, there would be no rain, snow, thunder or lightning and the sky would be very boring to look at. A cloud is a visible mass of tiny water droplets or ice crystals suspended in the air. Clouds can be thick or thin, big or little, and change form constantly.

They can be divided into four main types. Cumulus clouds are puffy masses that look like balls of cotton wool. Stratus clouds are flat and often cover the entire sky, extending many hundreds of square kilometres. Cirrus clouds are wispy and form up to 13km above the surface of the Earth. Dark clouds that bring rain are called nimbus. Cloud names can be combined. For example, cumulonimbus and nimbostratus are the names given to different clouds that produce rain.

Floating saucer
From a distance, this lenticular (lens-shaped) cloud looks rather like a flying saucer. These clouds often form in waves that develop downwind of a mountain range. Lenticular clouds are often elongated and usually have well-defined outlines. Frequently, they form one above the other like a stack of pancakes.

cloud of ice crystals

cloud of water droplets

water vapour rises

How clouds form
When warm humid air rises and cools, the vapour turns into droplets of water and forms clouds. If the air is very cold, the vapour will turn into a cloud of ice crystals.

Cirrocumulus clouds
The small ripples in cirrocumulus clouds look rather like the scales of a fish. The expression "mackerel sky" is used to describe a sky full of cirrocumulus clouds. These high white clouds are rounded and composed of ice crystals.

Cirrus clouds

The most common high clouds are the cirrus clouds, which are thin, wispy and made up of ice crystals. High winds can blow these clouds into long streamers called mares' tails.

Cumulus clouds

Puffy cumulus clouds take on a variety of shapes, but they most often look like balls of cotton wool. They have round tops and when they are dark and deep, they bring rain.

Cumulonimbus clouds

Cumulonimbus are thunder clouds. They are the largest clouds of all and form from cumulus clouds, often sprouting an anvil-shaped top, and produce heavy showers of precipitation.

cirrus

cirrostratus

cirrocumulus

altostratus

altocumulus

cumulonimbus

stratocumulus

cumulus

stratus

Clouds at different heights

Clouds can be grouped according to how high they are above the Earth's surface. High clouds include cirrus clouds. Altostratus and altocumulus are middle clouds. Stratus clouds are examples of low clouds.

The Sun's halo

A spectacular halo around the Sun is caused by a high cirrostratus cloud. Tiny ice particles in the cloud refract, or bend, light from the Sun to create a luminous ring.

RAIN AND DEW

IN SOME clouds, tiny water droplets remain suspended in the air. In other clouds, the droplets bump into one another and coalesce (join together). As the water droplets get larger and larger, they become too heavy to stay in the air. Eventually, the droplets fall out of the cloud as rain. Raindrops that reach the Earth's surface are rarely larger than 5mm across.

Rain is the commonest form of what meteorologists call precipitation. Precipitation is any form of water that falls from the atmosphere and reaches the ground. Dew is a form of precipitation. During a cold night, dew forms on surfaces such as leaves and the ground. The cold surfaces make the water vapour condense into droplets of liquid water.

From small beginnings
Rain falls from the dark base of a cumulonimbus cloud over the hills in the distance. Cumulonimbus clouds start as small, fluffy white cumulus clouds. Then they begin to grow and develop a dark base. Cumulus clouds sometimes mushroom into massive thunderclouds that reach up to 15km high and a severe thunderstorm may develop.

Walk in the rain
It is fun to go out in the rain but only if you dress properly. If you are wet, your body loses heat through the wet skin. If your body loses too much heat too quickly, a life-threatening condition called hypothermia may result.

A dewy web
Dew drops glisten on a spider's web. The dew formed on the web when water vapour in the air cooled during the night and condensed as water droplets.

Rain to come
Dark, stormy nimbus clouds are piling up in the sky near Majorca in the Mediterranean. The clouds are low and soon it will be raining hard.

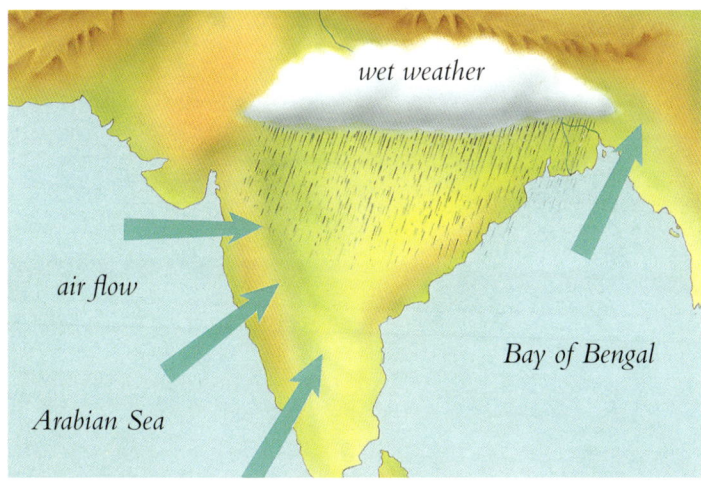

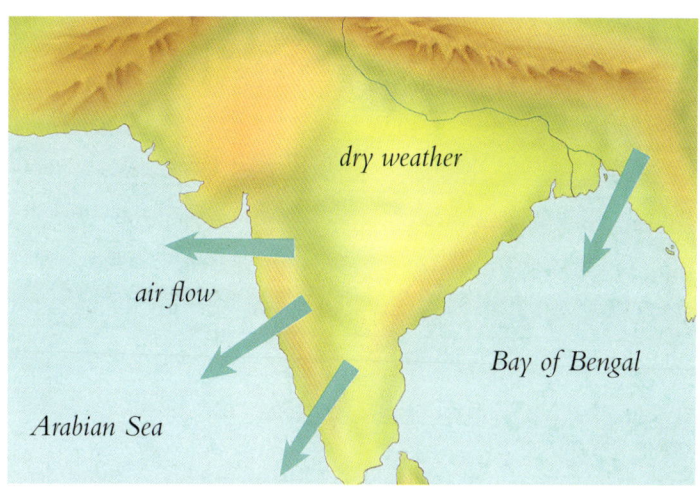

The summer monsoon

Some of the heaviest rain in the world falls in Asia during the monsoon period. During the summer, the air over the continent is warmer than the air over the water, and wind blows from the warm ocean. The winds bring heavy showers and thunderstorms, which can lead to flooding.

The winter monsoon

During the winter, the air over continental Asia becomes much colder than the air over the ocean. As a result, air flows out over the ocean. The winter monsoon provides southern Asia with generally fair weather and a dry season.

Railways into rivers

Days of almost continuous heavy summer monsoonal rains have turned this railway in Bangladesh into a river. The ground has become waterlogged, so the water will not drain away.

Moist air

In hot, tropical regions such as Hawaii, rain falls regularly throughout the year. Hawaii is surrounded by the Pacific Ocean. As a result, winds that blow from the sea to land bring air saturated (filled) with water vapour.

Summer burst

The monsoon rains fall across large areas of the tropics in summer, from northern Australia to the Caribbean. This monsoon rain is falling on a river in Indonesia.

MAKING RAINBOWS

RAINBOWS can often be seen during showery weather when the Sun is quite low in the sky. White sunlight is actually made up of a mixture of seven different colours – red, orange, yellow, green, blue, indigo and violet. Raindrops split up sunlight into a spread, or spectrum, of these separate colours to form a rainbow. The biggest rainbows form when the Sun is low in the sky, so they are most commonly seen in the evenings or mornings. They are also less common in the tropics, where the Sun is higher in the sky than in regions further north or south.

Other coloured effects can be seen in the sky. Sometimes a circle made up of faint rainbow colours, called a halo, forms around the Sun or the Moon. Ice crystals in front of the Sun or Moon split up the sunlight into a spectrum.

White light can be split up into a colourful rainbow spectrum by shining it through a prism (a triangular wedge of glass). In the experiment, you can produce your own rainbow by shining light through a "wedge" of water.

Midday rainbow
The mist of water created by waterfalls, such as Victoria Falls on the border of Zambia and Zimbabwe, creates the perfect conditions for the formation of rainbows.

Splitting up white
If you shine white light through a prism, it splits up into different colours and emerges from the other side as a rainbow – a coloured band called a spectrum.

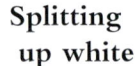

Split into seven
Rainbows are made up of seven different colours: red, orange, yellow, green, blue, indigo and violet. Here you can see a double rainbow.

FACT BOX

• White light is not really white. In fact, it is made up of seven different colours, as shown above. These colours mix together to make white light.

• When light travels from air into water or glass (or back the other way) it is refracted, which means it bends. Different colours in the light bend more than others. Blue light bends most, red light least. As a result, the colours start to separate out and the result is a spectrum – the colours of the rainbow.

PROJECT

SPLIT LIGHT INTO A RAINBOW

You will need: *mirror, dish, reusable adhesive, jug of water, torch, piece of white card.*

1 Carefully lean the mirror against the side of the dish. Use two small pieces of reusable adhesive to stick either side of the mirror to the dish at an angle.

2 Pour water into the dish until it is about 4cm in depth. As you fill the dish a wedge-shaped volume of water is created alongside the mirror.

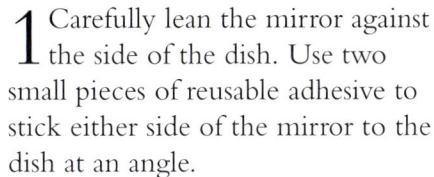

3 Switch on the torch. Shine the beam on to the surface of the water in front of the mirror. This should produce a spectrum or "rainbow".

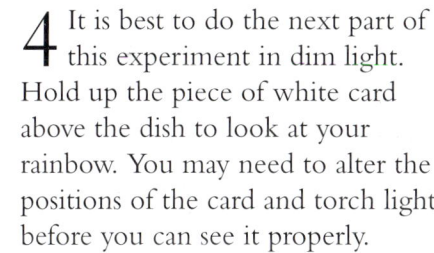

4 It is best to do the next part of this experiment in dim light. Hold up the piece of white card above the dish to look at your rainbow. You may need to alter the positions of the card and torch light before you can see it properly.

THUNDER AND LIGHTNING

Mother of lightning
Sieou-wen-ing is a mythological character from ancient China. She is thought to be the mother of lightning. In this picture, she is sending bolts of lightning towards the Earth.

A THUNDERSTORM occurs when warm, humid air rises. The upward air movement may be due to the uneven ground or temperature below. Huge, dark cumulonimbus thunderclouds develop overhead, flashes of lightning may fill the sky and the ground often trembles with a booming sound wave called thunder.

Lightning is a huge discharge of electricity. In a thundercloud, tiny drops of water and ice carry little bits of electricity, called electrical charges, which build up in parts of the cloud. In time, this charge becomes so great that electricity jumps to the ground or to other clouds, creating great sparks of lightning. The lightning heats up the air to a high temperature and makes it suddenly expand. This creates the explosion we hear as thunder.

Ice block
Hailstones are pieces of ice that fall from clouds. This one is 15cm across. Most are much smaller, but they become bigger the longer they stay up in the clouds.

Flashes of lightning
Lightning illuminates the night above the distant mountains. Light travels so fast that we see a flash of lightning almost instantly. The sound of thunder takes much longer to reach our ears, however, because sound waves travel more slowly than light. Sometimes, lightning is seen but no thunder is heard. This happens because during a thunderstorm, the air moves erratically and often scatters the sound waves. As a result, the thunder cannot be heard.

Lightning conductor

An experimental lightning conductor helps scientists study lightning. Electricity is easily conducted (passed along) through metal. Many tall buildings have a metal rod at the top. If lightning strikes, the metal conducts the electricity safely to the ground, and the building will not be damaged.

Struck by lightning

Lightning has struck a tree and left a trail of exposed wood in the bark. Lightning bolts often strike tall objects such as trees and buildings.

Bolt from the blue

A powerful streak of lightning discharges from thunderclouds and strikes a tree. The air surrounding lightning often rises to 30,000°C and may set the whole tree on fire.

CHARGING UP

THE lightning that flashes in the sky during a thunderstorm is not the same kind of electricity that makes your television or radio work. Lightning is a form of static electricity, which is made up of tiny electrical charges. These little bits of electricity can build up to create a much bigger charge, or voltage. Unlike ordinary electricity, the electrical charge does not usually flow away, which is why it is called static (not moving) electricity. In a thundercloud, the static electricity builds up so much that the air cannot hold it, and it jumps around as lightning flashes. In these experiments, you can build up small electrical charges by rubbing things together.

MAKE STATIC ELECTRICITY

You will need: *balloons, balloon pump, hairbrush.*

Van de Graaff generator
If you put your hands on a device called a Van de Graaff generator, your hair stands on end. A belt of material carries tiny electrical charges, which build up on a metal sphere, making static electricity.

1 Blow up a number of balloons with a balloon pump. Rub the balloons against a jumper or something made from wool. Put the balloons in different places.

2 Put the balloons on the ceiling, on the walls and even on your friends. The tiny electrical charges that form static electricity will make the balloons "stick" to things.

3 You can make your hair charge up with static electricity, too. Brush your hair when it is dry. Then hold the hairbrush near your hair. It will make your hair stand on end.

JUMPING ELECTRICITY

You will need: *plastic sheet, sticky tape, rubber gloves, metal dish, fork.*

1 Lay out the sheet of plastic on the table and secure the edges with sticky tape. This prevents the sheet from sliding around and disrupting your experiment.

2 Put the rubber glove on one hand. Slide the metal dish back and forth over the plastic sheet for a few minutes. This will charge the dish with static electricity.

Hands-on activity
If you put your hands over a plasma ball, little bolts of electrical charge move like lightning towards your hands. The "lightning" consists of harmless flashes of static electricity, travelling in a plasma (sea) of electrified particles.

3 Hold the fork with your ungloved hand. As you bring the fork close to the dish, you should see a spark jump. It is easier to see this in the dark.

FLOOD AND DROUGHT

Playtime
The flooded streets of Pahang, Malaysia, are a playground for some, but most people are mopping up the water after it has swept through their houses.

FLOODING (too much water) and drought (too little water) can have a devastating effect on the environment and the people who live in it. The people who live in these areas have to try to deal with these extreme conditions.

In some parts of the world, flooding occurs regularly in certain seasons, such as during the summer monsoons in India. Sometimes, flooding can be caused by cyclones. In February 2000, a cyclone with winds blowing at up to 260km/h caused widespread flooding in Mozambique. Tens of thousands of people were killed or made homeless as vast areas of land became submerged in water. Sudden flooding following heavy rains can be equally destructive. In December 1999, prolonged rain created terrible mudslides that buried or swept away whole towns and villages along the northern coast of Venezuela.

At the same time, east Africa was suffering from severe drought. Two years had passed without the usual seasonal rains. The crops had failed, livestock were dying in their thousands and much of the population were suffering from malnutrition.

Breaking records
A weather satellite image reveals the extent of the flooding of the Missouri and Mississippi rivers in the USA in July 1993. The city of St Louis is coloured purple at the bottom of the picture. In the summer of 1993, the American Midwest suffered the worst flooding since records began. River levels rose up to 15m above normal.

A dangerous delta
A Bangladeshi family stands by the remains of their home after a devastating storm in the Ganges Delta. A cyclone has surged inland from the Indian Ocean, crossed the flat Delta region, and flattened everything in its path.

Around the waterhole

Animals gather around a waterhole during the dry season on the east African savanna. During the dry season no rain falls at all. All the water dries up, and the vegetation available for the animals to graze on gets scarcer. Only a few waterholes are left, but even these will shrink as the Sun relentlessly beats down on the plains.

Death in the drought

In 1992, parts of east Africa suffered one of the worst droughts ever recorded. Crops were devastated, and livestock such as cattle were killed by the thousand.

Dry and lifeless

These trees have died through lack of water. Drought is caused by a lack of precipitation, but it can also be caused by hot, dry winds and frequent fires. These elements combine to take moisture from the soil and use up groundwater beneath the top levels of the soil.

Aid for Ethiopia

Families from Ethiopia in east Africa gather at an aid centre to collect food. Ethiopia is one of the world's poorest countries. Most people live by farming the land, but drought results in a very poor harvest and widespread famine. In 1984, more than 800,000 people are known to have died in one of the worst droughts seen to date.

GAUGING THE RAIN

Keeping dry
Umbrellas were probably invented in China as early as the 2nd century BC.

THE amount of precipitation (rain, sleet, snow or dew) that falls from the atmosphere varies widely throughout the world. Heavy rain falls regularly in tropical regions around the Equator. Here, the air contains plenty of moisture evaporated from the warm oceans. The summer monsoon rains that occur over southern Asia can reach record amounts. Cherrapunji in north-eastern India receives an average of 10m of rainfall each year, most of which falls during the summer monsoon between April and October. These rains are essential to the agriculture of southern Asia. Since so many people depend on the monsoon to survive, meteorologists need to predict how much rain will fall so that food crops will grow.

How much rainfall do you get where you live? Make this simple rain gauge to measure the amount. Meteorologists use a similar rain gauge at many weather stations around the world. If it is very rainy where you live, this project will keep you busy, but if you live in a desert region, you may have to wait a long time for any rain!

MEASURING RAINFALL

You will need: *scissors, sticky tape, large jar (such as a sweet jar), ruler, ballpoint pen, large plastic funnel, tall narrow jar or bottle, notebook.*

1 Cut a piece of tape to the height of the jar and stick it on. Mark a scale on the tape at 1cm intervals. Measure the diameter of the jar and cut the funnel to the same size.

2 Place the funnel in the jar. Put the gauge outside in an open space away from any trees. Look at the gauge at the same time each day. Has it rained in the last 24 hours?

3 If it has rained, use the scale to see how much water is in the jar. This is the rainfall for the past 24 hours. Make a note of the reading. Empty the jar before you return it to its place.

PROJECT

Being more precise

Measure rainfall more accurately by using a separate narrow measuring jar. Cut a length of tape to the height of the narrow jar and stick it to the side. Pour some water into the large collecting jar up to the 1cm mark. Now pour this water into the smaller measuring bottle. Mark 1cm where the water level reaches. Divide the length from the bottom of the bottle to the 1cm mark into 10 equal parts. Each mark you make will be equivalent to 1mm of rainfall. Extend the scale past the 1cm mark to the top of the measuring bottle.

All-in-one weather instrument

You can buy an all-in-one weather instrument, which measures temperature, rainfall, wind direction and wind speed. These are handy if you don't have much space to set up lots of equipment.

Automatic weather station

Meteorologists use all-in-one weather instruments to measure different features of the weather. These instruments automatically monitor local weather conditions, such as wind speed and direction, air pressure, temperature, humidity and solar radiation.

FACT BOX

• The wettest place in the world is Mawsynram in India. Here, an average of nearly 12m of rain falls every year.

• New York and Sydney have a little over 1m of rain a year. Paris and London have about 60cm a year.

• The driest place in the world is the Atacama Desert, west of the Andes in Chile, South America. In parts of the desert just a few showers fall every 100 years.

• Sea storms can cause worse flooding than rainfall. The large waves that form can submerge coastal areas.

SNOW AND ICE

SNOW falls in winter in many countries. It also falls all year round in places near the North and South poles and at the top of mountains. Most of the precipitation that actually reaches the ground starts as snow. At the top of high clouds the temperature is below the freezing point of water, and ice crystals form. Ice crystals join together to form snowflakes that fall from the clouds when the snowflakes become too heavy. If the lower air is warm, however, the snowflakes will melt and turn into rain. A mixture of snow and rain sometimes falls as sleet. This occurs when the falling snowflakes start to melt and then turn back into ice as they pass through a freezing layer of air.

On many winter nights the ground becomes snow-white even when it has not been snowing. This white covering is called frost, which forms when the ground gets cold and water vapour in the air condenses on it. The water immediately freezes into tiny sparkling crystals of ice.

Works of art
Snowflakes are made up of masses of tiny ice crystals. Under a microscope, the most common snowflake form is a branching star shape called a dendrite.

Death in the valley
In February 1999, an avalanche devastated a village in the Chamonix Valley in the French Alps. About 40,000 tonnes of snow hurtled down the mountain slopes at 200km/h. The snow buried houses and cars, killing 12 people.

Avalanche!
Thousands of tonnes of snow break loose in an avalanche on Mount Everest in the Himalayas. Avalanches career downhill at tremendous speeds, destroying everything in their path. They occur when the weight of the snow on the mountain exceeds the forces of gravity and friction that hold the snow in place.

Ice on glass

Ice crystals have formed on a windowpane. If glass gets very cold, water vapour in the air condenses on it, forming crystals where it freezes.

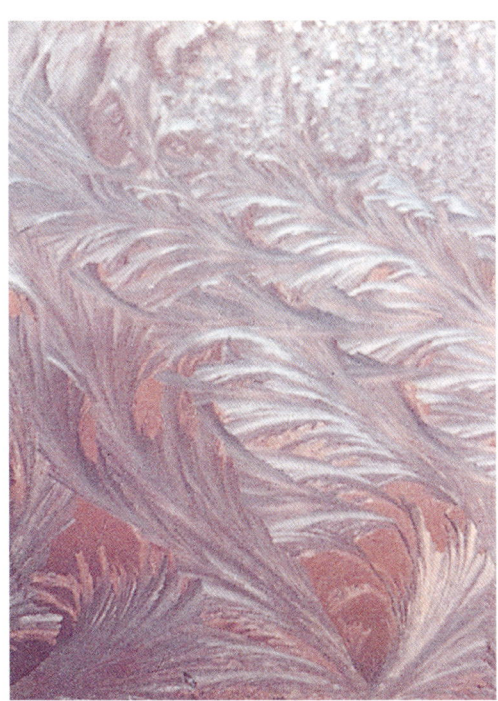

Jack Frost

Jack Frost is a mythical character who is thought to make the beautiful icy patterns you find outside on trees, plants and fences during cold weather.

Chunks of ice

Most of the ice in the world is found in the ice caps in the Arctic and Antarctic. Around 6 million sq km of ice cover the Arctic Ocean in the north. This huge ice pack is broken into large floes (sea ice) by the wind and ocean currents. Antarctica is covered by a permanent ice cap, which is over 3km thick at the centre of the continent.

Ice storms in Quebec

During the winter of 1998, the province of Quebec in Canada suffered a storm of icy rain that lasted for a whole week. Ice up to 8cm thick collected on trees, electricity pylons and cables. The ice was so heavy that trees and pylons collapsed, resulting in widespread damage and disruption. The more remote areas of Quebec did not have any electricity for a month. Many farmers lost livestock because they could no longer feed them or keep them warm. This was the worst ice storm experienced in this region for at least 100 years.

COPING WITH COLD

Needle leaves
The needle-like leaves of these conifers in the Rocky Mountains of North America lose much less heat than broad leaves would.

Blubber is best
Seals spend a lot of time swimming in the ice-cold waters of the Arctic and Antarctic oceans. Seals are warm-blooded, which means they can regulate their own body heat. Under their skin, a fat-filled layer of spongy tissue, called blubber, insulates their bodies from the cold.

Penguin playground
Adelie penguins gather on the ice along the coast of Antarctica. The temperatures can drop to as low as −90°C in Antarctica. Penguins are protected from the severe cold by their closely packed, oily feathers and an insulating layer of fat under their skin.

ONLY a tiny proportion of animals and plants live in the cold temperate regions in the far north of North America, Europe and Asia and in the polar regions around the North and South poles. In these areas the winters are long, and temperatures often fall below −50°C. The plants and animals that live in these bitterly cold parts of the world are well adapted to the environment. The main plants are evergreen conifer trees. These form a great northern, or boreal, forest region that spans the continents of North America, Europe and Asia.

The largest animals of the boreal forests are caribou and moose (elk). These animals have thick fur and can survive on almost any type of vegetation. They shelter in the forests in the winter but venture on to open ground farther north in summer. This open ground is called the tundra, which is covered by snow and ice for most of the year. A few metres below the surface, the ground is permanently frozen. It is too cold for trees to grow, so grasses and low shrubs make up the vegetation. These plants make the most of the short summer months by growing, flowering and seeding rapidly.

Farther north still, on the permanently frozen north polar ice cap, plants do not grow. Polar bears hunt seals that swim in the icy waters of the Arctic Ocean. Polar bears have thick fur with a layer of fat underneath to protect their bodies against the cold. Seals and whales have a thick layer of blubber beneath the skin to insulate them from the cold. Seals and whales are also found around Antarctica at the opposite end of the Earth.

Saami in the snow

A caribou herder from Kataukeino in northern Norway relaxes on his snow scooter during the spring migration of caribou. He is one of the Saami people from northern Scandinavia, and is dressed in traditional clothing – a parka and trousers made out of animal skins and trimmed with fur. These warm clothes will insulate him against the bitter cold of the Scandinavian tundra.

Fruit of the tundra

Arctic plants flower when the ground is barely free of snow, taking advantage of the short summer. The ground over the permanently frozen layer of soil, called permafrost, thaws in the summer. The plants then get a roothold. They cannot develop deep roots because of the solid layer of permafrost.

Like father, like son

An Inuit father and his son live in the Northwest Territories of Canada, which has one of the coldest climates on the Earth. Their bodies are well adapted to the climate. They are short, stocky and their faces have grown used to the cold. Inuit people have a very fatty diet to build up a thick layer of fat.

The treeless tundra

It is summer and the lower slopes of the mountains in Denali National Park, Alaska, USA, have lost their covering of snow. Although the ground just below the surface is still frozen, plants such as grasses and low shrubs can still survive.

ICE AGE OR GREENHOUSE?

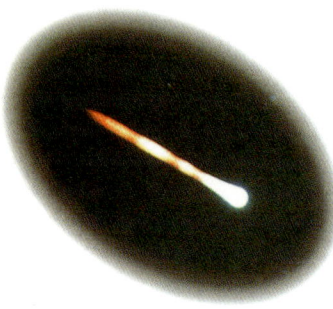

Ball of fire
A shooting star, or meteor, streaks through the atmosphere. Shooting stars are streaks of light created when lumps of rock and dust hurtle towards the Earth and burn up in the Earth's atmosphere. Parts of a rock often fall to the ground as meteorites. If a large meteorite hits the Earth enough material can be thrown into the atmosphere to bring about a change in climate. This may have caused the extinction of the dinosaurs some 65 million years ago.

THROUGHOUT the history of the Earth, the climate has undergone many changes. Over the past million years or so, the climate has alternated between periods of warmth and cold. During the cold periods, called ice ages, most of North America and Europe became covered in vast sheets of ice. The last cold period lasted until about 25,000 years ago. Scientists used to think that another ice age was approaching. In fact, the atmosphere has been warming at a dramatic rate since the 1970s. This may be due to a number of reasons.

Most importantly, humans are burning more and more fuels, such as coal and oil, which creates carbon dioxide. This gas is building up in the atmosphere, trapping the Sun's heat. As a result, the atmosphere acts rather like a greenhouse. As more and more carbon dioxide builds up in the atmosphere, this so-called "greenhouse effect" warms the climate. Scientists think that if humans continue to produce carbon dioxide by burning fossil fuels, the Earth's climate will warm by several degrees in the next 50 years. This may make the polar ice caps melt, causing sea levels to rise and flooding many countries. Rising temperatures also alter wind patterns and ocean currents, disrupting weather patterns and climates throughout the world.

Holding back the tides
The Thames Flood Barrier at Woolwich in London will prevent the city from being flooded in case of an exceptionally high tide. If the greenhouse effect becomes a reality, this preventative measure may save millions of people who live in the city.

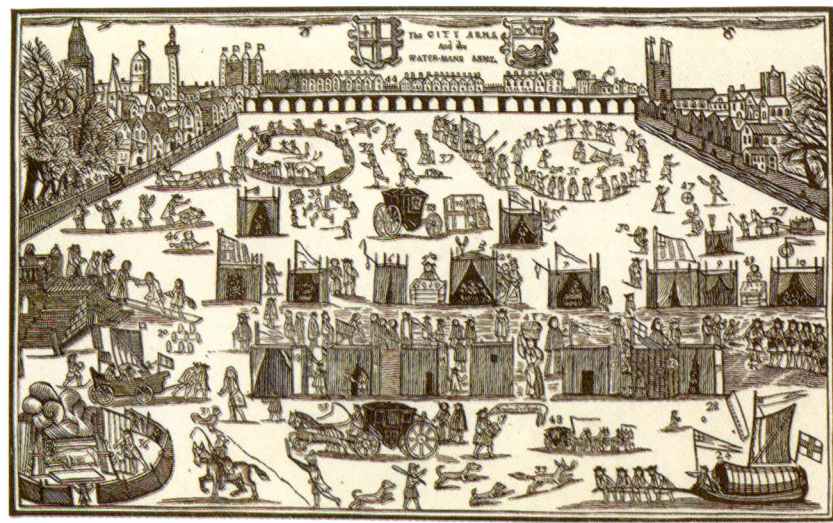

Frozen over
A change in the heat output of the Sun may have caused the Little Ice Age that occurred in the 1600s. During this time, the River Thames in London froze so hard that a series of frost fairs could be held on it. The picture above shows the Frost Fair that was held in 1683.

The ash clouds of Pinatubo

After laying dormant (sleeping) for more than 500 years, the volcano Mount Pinatubo erupted in June 1991. As the volcano erupted, it threw vast clouds of ash high into the air, blocking light from the Sun for days. Torrential rains followed the eruption and caused mud and ash slides, which devastated the surrounding countryside. Ash particles in the atmosphere were also responsible for cooler summers around the world for several years.

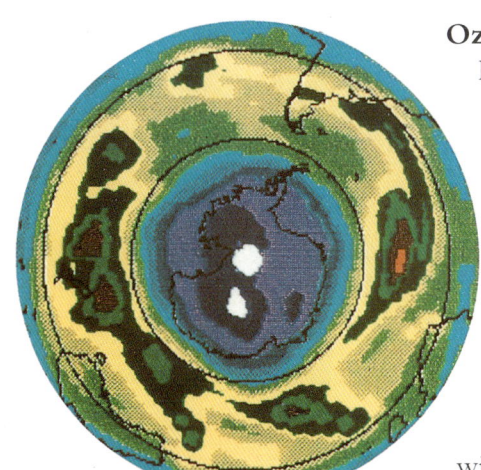

Ozone hole

In the 1980s, scientists discovered that the ozone layer, part of the Earth's atmosphere, was thinning over Antarctica. If the ozone layer becomes too thin, it will let more rays through from the Sun, which will harm the Earth.

Cycle safely

A cyclist wears a protective face mask to filter out the harmful fumes of busy city traffic. Gases released by car engines are now known to be responsible, in part, for the greenhouse effect.

SOHO image

The SOlar Heliospheric Observatory (SOHO) space probe took this picture of the scorching surface of the Sun. Using space observatories such as these, meteorologists learn more about how and why changes take place in the Sun's energy output.

STUDYING THE WEATHER

METEOROLOGISTS do two main jobs. They collect data and process information about the weather from day to day. Then they use this information to help them forecast future weather trends. Meteorologists collect information from weather stations scattered all over the world. They use a variety of measuring instruments. Thermometers measure the temperature, barometers measure air pressure and hygrometers measure the humidity of the air. Weather vanes show the direction of the wind and anemometers measure its speed.

Increasingly meteorologists are turning to space technology to help them. They send weather satellites into orbit to take pictures of clouds and measure weather conditions in the air. Satellites are useful because they can record weather in remote regions where there are no weather stations.

Blast-off
A rocket blasts off from Cape Canaverel, Florida, in the USA. The rocket is carrying a weather satellite that will orbit the Earth and send images to meteorologists back on Earth, along with other weather data.

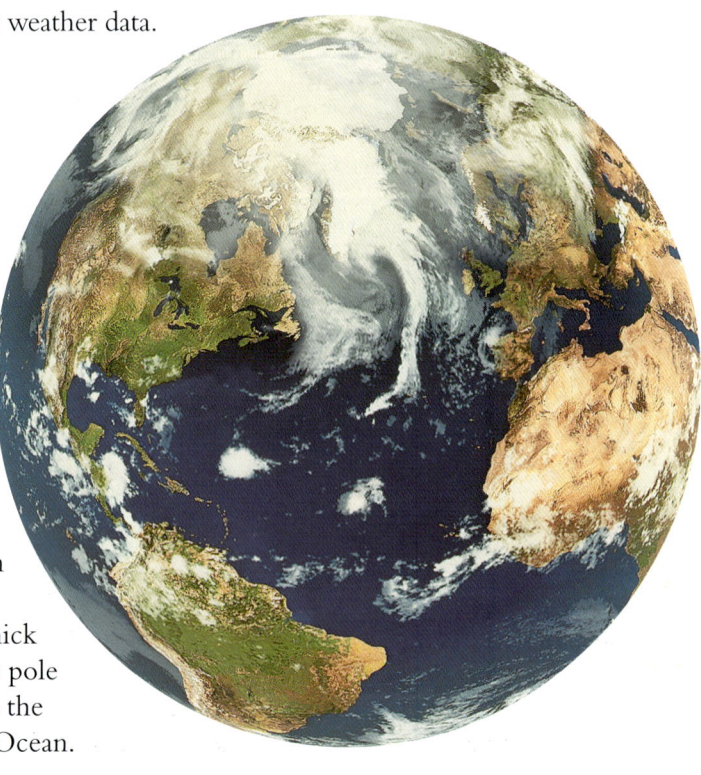

Cloud in the north
A number of satellite pictures have been joined together to give a true colour image of the Earth. The polar ice cap at the North Pole is shown at the top here. Thick cloud covers the pole and reaches into the North Atlantic Ocean.

Ready for launch
A meteorologist launches a weather balloon at the Kourou Space Centre in South America. The weather balloon will be tracked once it is in the air. The direction it travels will indicate the wind's direction. Readings of temperature and humidity from the instruments it carries will also be sent back.

Measuring sunshine

A sunshine recorder is made up of a glass ball that acts rather like a lens. When the Sun shines, the glass ball focuses the light on to a sheet of paper below the ball. The focused light leaves a scorch mark on the paper that is matched to a calibrated time scale. In this way, the meteorologist can record how long the Sun shines each day.

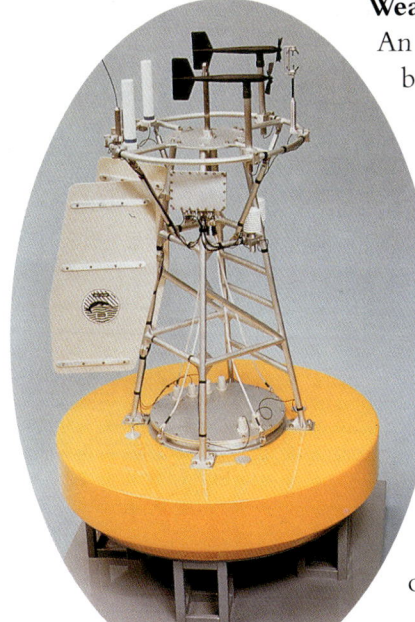

Weather at sea

An automatic weather buoy carries many types of weather-recording devices, such as anemometers, barometers, hygrometers and thermometers. The readings from the different instruments are transmitted by radio to weather stations or passing satellites.

Head for the clouds

A pilot flies his plane straight into a storm. He works for the National Weather Service in the USA. The plane is a specially strengthened aircraft that can cope with violent air currents, lightning and bombardment by hailstones. Instruments beneath the wings monitor weather conditions.

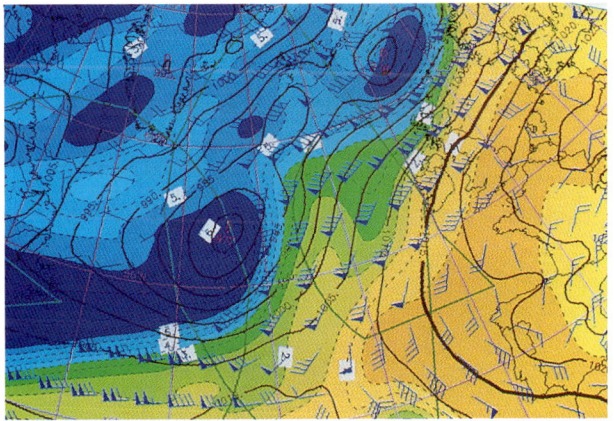

Storm map

A computer map is used to predict the likelihood of a storm in the Atlantic Ocean. The solid lines are called isobars and link regions of equal air pressure. Storms often occur in regions where the isobars are close together.

Computer forecasting

Meteorologists use computers to predict how the weather will change. This can help them make more accurate forecasts.

YOUR WEATHER STATION

METEOROLOGISTS work at about 12,000 weather stations worldwide, gathering information about the weather. They feed the data they gather from satellites, balloons, weather buoys and other instruments into powerful computers to provide them with an overall view of the weather and how it may change. Using this information, they can draw weather maps that show the state of the weather at any one time, using symbols to represent conditions such as rainfall, wind direction and pressure. They also use this information to draw other charts that they use to make a forecast of the weather.

You can set up your own weather station to record weather conditions with a few simple instruments. You will be able to use some of the instruments you have made in earlier projects, such as the weather vane, hygrometer and rain gauge. In addition, you will need to buy a thermometer and a barometer (which measures air pressure), both of which can be bought fairly cheaply.

How hot is it?
A thermometer must always be kept in the shade to measure the air temperature accurately. If the device is left in direct sunlight, the liquid will also absorb energy from the Sun. As a result, the thermometer will indicate a temperature higher than the actual air temperature.

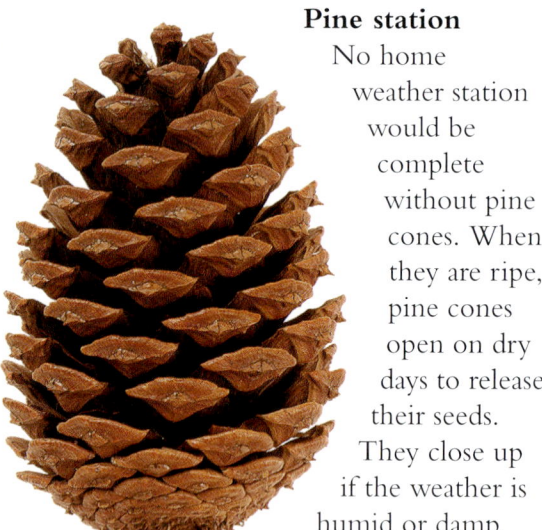

Pine station
No home weather station would be complete without pine cones. When they are ripe, pine cones open on dry days to release their seeds. They close up if the weather is humid or damp.

Make a note
Take measurements with your weather instruments every day. Write them down in a special weather book. Also, make a note of what the weather is like generally – fine, cloudy, drizzle, frosty and so on. Don't forget to make a note of the date!

How much did it rain?
Your rain gauge will tell you this. Measure the amount of water in the jar. Use the measuring bottle to be more accurate. Always empty the jar when you have finished.

Dry or damp

Seaweed is a useful item to have in your weather station. Like pine cones, seaweed changes as the humidity changes. If the weather is dry, the seaweed feels dry and brittle. If the weather is humid, however, the seaweed feels flexible and damp.

Cumulus clouds

As well as using the instruments you have made to predict what the weather will be like, you can make general predictions, too. Studying the clouds is often a good way of telling what sort of weather is in store. Puffy cumulus clouds that grow in size and turn darker suggest that there could soon be showers. However, these clouds are scattered, so showers will not last for long.

Red skies

Red clouds at dawn is often a sign that rain is on the way. A red sky at night, however, can be a good sign, promising that the next day will be fine. An old saying sums this up with the words "Red sky at night, shepherd's delight. Red sky in the morning, shepherd's warning."

How humid is it?

Your hygrometer will help. Note the position of the pointer on the scale. When the pointer tilts up, the air is moist and rain could be on the way.

Which way is the wind blowing?

Remember that the arrow on your weather vane points in the direction from which the wind is blowing. So a north wind blows from the north.

GLOSSARY

acid
A class of chemical compounds which contains the element hydrogen combined with other elements. The hydrogen is split from the other elements when the acid is dissolved in water.

acid rain
All rain is slightly acidic, but when pollution by sulphur dioxide and nitrogen dioxide reacts in sunlight with oxygen and moisture in the sir, it creates acid rain.

amber
A pale brown or yellow, transparent semi-precious stone which is the fossilised remains of tree resin. It sometimes contains beautifully preserved prehistoric insects that became trapped in the sticky resin.

anemometer
An instrument used in weather forecasting to measure the speed and force of the wind.

anthracite
A variety of hard, shiny coal that burns with hardly any smoke and gives more heat than any other kind of coal.

archipelago
A large group of islands.

arete
A sharp, knife-edge ridge between two mountain glaciers.

atmosphere
The layer of air that surrounds a planet and is held to it by the planet's gravity.

atmospheric pressure
A measure of the weight of the atmosphere.

atoll
A ring-shaped coral island created as the reef round a volcanic peak goes on growing at sea level while the volcano sinks beneath the waves.

atom
The smallest part of an element that can exist. It is made up from many other smaller particles including electrons, neutrons and protons.

aurora
Spectacular displays of coloured lights in the night sky above the North and South poles.

barometer
An instrument used in weather forecasting to measure the pressure of the atmosphere.

biome
A community of plants and animals adapted to similar conditions over large regions of the world. Sometimes called major life zones.

calcite
One of the commonest minerals and the main constituent of limestone. Its chemical name is calcium carbonate.

carbon dioxide
A colourless, odourless gas containing the elements carbon and oxygen that is a part of the air people breathe. It is produced when fuel containing carbon is burned in the air.

cirque
A deep hollow in the mountains carved out by the head of a glacier.

climate
The typical weather pattern of place during the year.

collision zone
A zone on the Earth's surface where the edge of one tectonic plate is being forced into, and under, another. It is marked by lines of volcanoes and strongly distorted rock layers.

compass
An instrument, containing a magnetised strip of metal, used for finding the direction of the Earth's magnetic north.

condensation
The process by which water vapour becomes a liquid.

constructive boundary
The edge of one of the Earth's plates, where new plate material is forming.

continental crust
The thick part of the Earth's crust under continents.

continental drift
The generally accepted theory that continents move slowly around the world.

continental shelf
The zone of shallow water in the oceans around the edge of continents.

convection
The rising of hot air or fluid, caused by the fact that it is lighter than its surroundings.

core
The dense, hot metallic centre of the Earth.

crevasse
A deep crack, typically in the surface of a glacier.

crust
The solid outer shell of the Earth, varying from 5-80km thick.

cyclone
An area of low pressure into which winds spiral clockwise in the Southern Hemisphere and counter-clockwise in the Northern Hemisphere.

denudation
The gradual wearing away of the landscape.

destructive boundary
A region of the Earth's crust where one of the plates of the crust is colliding with another and being destroyed.

dew
Water that condenses onto objects near the ground when their temperatures have fallen below the surrounding air.

drought
A long time without rain, when living things do not have the water they need.

dust devil
A small, rapidly rotating wind that is visible due to the dust it picks up from the ground as it rotates.

earthquake
An often violent shaking of the Earth's crust, caused when the plates in the crust try to slide past or over each other.

ecosystem
A community of living things interacting with each other and their surroundings.

element
A chemical substance, such as gold, carbon and sulphur, that cannot be further broken down into other elements.

epicentre
The region on the Earth's surface that lies directly above the focus of an earthquake.

Equator
The imaginary circle around the middle of the Earth between the North and South poles. This region has a climate that stays hot all year round.

era
A major subdivision of geological time.

erosion
The gradual wearing away of the land by weathering and agents of erosion such as rivers, glaciers, wind and waves.

fault
A fracture in rock where one block of rock slides past another.

fossil
The remains, found preserved in rock, of a living creature that lived in the past.

front
The point or boundary where two air masses that have different temperatures and different amounts of moisture meet.

fumarole
An opening in the ground in volcanic regions, where steam and gases can escape.

glacier
A river of solid ice.

geology
The scientific study of the Earth and the changes the changes that take place on its surface and in the rocks below.

geophysics
A sister science to geology concerned with the physical properties of rocks such as magnetism, density and radioactivity.

geothermal energy
The energy created in areas of volcanic activity by the heating of rocks below the Earth's surface.

geyser
A fountain of steam and water that spurts out of vents in the ground in volcanic regions.

glacial drift
All the material deposited by a glacier or ice sheet.

glaciation
The moulding of the landscape by glaciers and ice sheets.

gneiss
A common type of metamorphic rock produced deep in the Earth's crust. It is formed when other rocks are subjected to strong pressures and high temperatures.

greenhouse effect
The way certain gases in the atmosphere trap the Sun's heat like the panes of glass in a greenhouse.

groundwater
Water which percolates downward from the surface of the Earth into spaces in the rocks below.

Gulf Stream
A warm, swift, narrow ocean current flowing along the east coast of the USA and towards western Europe.

guyot
A flat-topped mountain under the sea, typically a volcano that has been eroded at the summit by waves, and which has then been submerged.

hanging valley
A side valley cut off and left hanging by a glacier.

hot spot
A place where plumes of molten rock in the Earth's mantle burn through the Earth's crust to create volcanoes.

hot springs
Places in volcanic regions where water has been heated underground by rocks bubbles to the surface.

humidity
A measure of the amount of water, or moisture, in the air.

hydrocarbon
A chemical compound containing the elements hydrogen and carbon.

ice age
A long cold period when huge areas of the Earth are covered by ice sheets.

igneous rock
A rock that forms when magma (hot, molten rock) cools and becomes solid. One of three main types of rock, created as hot molten rock from the Earth's interior cools and solidifies.

intrusive rock
A rock that forms underground when hot molten rock forces its way into existing rock layers and then cools.

iron oxide
A compound (mixture of elements) found all over the Earth that contains both the elements iron and oxygen.

lava
Hot molten rock emerging through volcanoes, known as magma when underground.

limestone
A rock, usually sedimentary, formed almost entirely of the mineral calcite.

lithosphere
The rigid outer shell of the Earth, including the crust and the rigid upper part of the mantle.

magnetism
An invisible force found in some elements but especially in iron, which causes other pieces of iron to be either pushed apart or drawn together.

mantle
The very deep layer of rock that lies underneath the Earth's crust.

metamorphic rock
Rock created by the alteration of other rocks by heat or pressure.

meteor
A piece of rocky material from space which burns as it falls through the Earth's atmosphere producing a streak of light.

meteorologist
A person who studies the science of meteorology and forecasts and reports on the weather.

meteorology
A science that studies the atmosphere, climates and weather conditions in regions throughout the world.

mica
A common crystalline mineral, found in igneous rocks, which splits into thin, flexible, transparent sheets.

mid-ocean ridge
A long jagged ridge on the sea floor along the gap between two tectonic plates which are moving apart.

mineral
A naturally occurring substance, found in rocks.

monsoon
A wind that reverses its direction in winter and summer, commonly affecting southern Asia around the Indian Ocean and often bringing heavy rains in the summer season.

moraine
Sand and gravel deposited in piles by a glacier or ice sheet.

ore
A mineral from which a useful material, especially metal, is extracted.

ozone
A form of oxygen gas. It is poisonous, but a high layer in the stratosphere protects us from the Sun's harmful ultraviolet radiation. The ozone hole is where the ozone in the stratosphere is very sparse.

palaeontology
The study of fossils.

permafrost
An area of ground that has remained frozen for at least a year, and usually much longer.

planet
One of the nine large bodies in the Solar System that circles around the Sun. The Earth is one of the nine planets.

precipitation
Any form of water (rain, hail, sleet or snow) that comes out of the air and falls to the ground.

recycle
To convert something old into something new. In nature, old rocks are continuously being changed into new rocks by the movement of the Earth.

Richter scale
A scale for measuring the strength of Earthquakes, devised by the US scientist Charles Richter.

rift valley
A valley formed when a strip of land drops between two faults.

savanna
A huge region of grassland typically found in a tropical climate.

sedimentary rock
A rock made up of mineral particles that have been carried by wind or running water to accumulate in layers elsewhere, most commonly on the beds of lakes or in the seas and oceans.

seismology
The study of the waves that earthquakes send out.

soil
Material produced from rock, at the surface of the Earth, by the action of the weather, plants and animals.

Solar System
The family of planets, moons and other bodies that orbit round the Sun.

strata
Layers of sedimentary rock.

subduction
The bending down of a tectonic plate beneath another as they collide.

tectonic plate
The 20 or so giant slabs of rock that make up the Earth's surface.

thermocouple
A thermometer scientists use to measure very high temperatures.

thermometer
An instrument for measuring temperature.

tidal wave
A huge ocean wave caused when an earthquake takes place on the seabed. It has nothing to do with tides.

till
Mixture of rock debris left by an ice sheet over a wide area as it retreats.

tor
A clump of big blocks of bare rock on top of a smooth hilltop.

tornado
An intense, rapidly rotating column of air that extends from a thundercloud in the shape of a funnel.

transpiration
The process in the water cycle by which plants release water vapour into the air.

tremor
A shaking of the ground.

trench
A valley in the seabed, marking the region where one plate of the Earth's crust meets another and is forced down into the crust.

tropics
Part of the Earth's surface that is between the Tropic of Cancer (at a latitude of 23.5 degrees north of the Equator) and the Tropic of Capricorn (at a latitude of 23.5 degrees south of the Equator).

tundra
A huge treeless area in the Arctic region where the ground beneath the surface is always frozen, even in summer.

typhoon
A hurricane that forms over the western Pacific Ocean.

vent
An opening in the ground.

volcanic bomb
A lump of molten material flung into the air from a volcano.

volcano
An opening in the Earth's crust from which molten rock escapes.

waterspout
A column of rotating wind over water that has characteristics of a dust devil and a tornado.

weather
The condition of the atmosphere at any particular time and place.

weathering
The breakdown of rock when exposed to the weather.

wind
Air moving in relation to the Earth's surface.

INDEX

ACKNOWLEDGEMENTS

PICTURE CREDITS
b= bottom, t= top, c= centre, l= left,
r= right

COVER
Front: Papillo, tc. Planet Earth Pictures, l, tr.
Telegraph Colour Library, br.
Back: Planet Earth Pictures, cl. Science Photo
Library, tr.

Bryan & Cherry Alexander: 174bl. Ardea:
145tl; /F Gohier page 143b /A Warren page
163cl. The Ancient Art & Architecture
Collection: 218c. The Art Archive: 118bl,
124br & 192tl. BBC Natural History Unit:
21tl, 24c, 30tr, 32br, 37c & bl, 39c, 43t & c,
45t & b, 53t & c, 58tl; /C Buxton: 160bl /B
Davidson: 161br /Pete Oxford: 126cr /Doug
Wechsler: 201cl. Biofotos: /B Rogers: 160br
/S Summerhays: 134tr, 153 & 182br.
Bridgeman Art Library: 120tr, 124cr, 125br &
167tr. British Antarctic Survey: 93cr, 99br &
c. British Atlantic Survey: 92t. British
Geological Survey: 87cr, 93tl, 105bl, 114br.
Bruce Coleman Collection Ltd: 27br, 36br,
55c, 58c, 65tl, 133br, 196bl, 240cl; /C
Atlantide page 148br /S Bond: 81tr /F
Bruemmer page 161c /J Cancalosi: 109bl /J
Cowan: 103bl /D Croucher: 81bl /G Cubitt:
141cr /J. Foott: 81tl /C Fredriksson: 104b /M
Freeman page 186t & 231br /Tore Hagman:
225cl /G Harris: 103br /Johnny Johnson:
241tr /Dr Scott Nielsen: 224br, 225tl & br
/Mary Plage: 227br /John Shaw: 124tl &
199bl /Kim Taylor: 226cr. Cephas: /M Rock:
161bl. Thomas Chatham: 75br. Corbis
Images: 19bl, 22tl, 23c, 55t, 56tl, 58c. Sylvia
Cordaiy Photo Library: 51c. De Beers: 70tr,
73cr, 124bl. Digital: 118br. Ecoscene: 28c.
Mary Evans Picture Library: 230tl, 242bl; /A
Rackham: 239tr. Frank Lane Picture Agency:
112cl, 114bl ; /S Ardito: 179tr /L Batten: 97tl
& 103tl /C Carvalho: 86br /S Jonasson page
137cr /S McCutcheon: 86bl, 89bl, 159bl &
175cl /C Mullen: 120bl /M Newman: 84tr
/M Niman: 80tl /M J Thomas: 81br /R
Tidman: 85br /T Wharton: 105tr /W
Wisniewski: 104tl & 114tr /M Withers:
153bl. Genesis Space Photo Library: 11tl,
18br, 179tl. GeoScience Features Picture
Library: 74cl, 76bl, 80tr, 91cl, 97cl & bl, 152tl
& 220tl. Getty One Stone: 9 t, 10 tl, 13 t, 15
c, 17 bl, 28 tr, 35 tl, 43 tl, 47b, 115bl & br,
120br, 191bl, 193bl, 198tl, 202bl, 204cr,
209cl, 238tr, 239cl, 240br, 244bl, 253br;
/Glen Allison: 202cr /David Austen: 197br
/John Beatty: 235cl /Peter Cade: 209tr /J F
Causse: 205br /Chris Cheadle: 203tl & 232tr
/Darrell Gulin: 221b /Paul Kenward: 204tl
/Laurence Dutton: 243br /David Hiser: 241bl
/Jerry Kobalenko: 212bl /Hiroyuki
Matsumoto: 193tr /Alan Moller: 191br &
213tr /Ian Murphy: 228tr /Frank Oberle:

191tr /Martin Puddy: 234tl /James Randklev:
221cr /Peter Rauter: 212br /Jurgen Reisch:
202tr /Lorne Resnick: 203bl /Manoj Shah:
235tl /Robin Smith: 193br /Brian Stablyk:
203br /Bill Staley: 226cl /Vince Streano:
201tr /Michael Townsend: 241tl /Larry
Ulrich: 191tl /John Warden: 221tl /Art
Wolfe: 203tr /Darrel Wong: 205tl. Robert
Harding Picture Library: 99bl, 102bl, 131cl;
/V Englebert: 94bl & br /N Francis: 118bl /R
Frerck: 175bl /P Hadley: 112br /L Murray:
97tr /Tony Waltham: 68-69, 115cl. Michael
Holford: 142tl & 180t. Hulton-Getty Picture
Collection: 130tr, 208cr & 214c. Image
Select: 148bl; /Caltech: 178b. JS Library
International: 181bl; /G Tonsich: 151cr.
Robin Kerrod: 194br, 205tr, 209cr, 222tl,
226bl, 227cl, 228cr, 239tl & 240tl. Landform
Slides: 155cr. Microscopix: 75bl, 93tr, 105cr
& br; /A Syred: 93bl. Milepost 9 1/2: 115tl.
Mountain Camera: /C Monteath: 179br /J
Cleare: 174t /T Kajiyama: 171br, 176t &
177cr. NASA: 85tr, 109br, 122tl & cl, 123 tl,
cl & cr. Natural History Museum: 71cr, 87tr,
114l & c. Natural History Photographic
Agency: 7cl, 21t, 27cl, 29t, 32tr, 33c, 38tl,
45c, 47c, 54tl, 57c, 59c, 60t, 62tr, 63c, 110c,
188-189, 251tr, 254tl, 255; /B&C Alexander:
55br /G Bernard: 91cr /D Woodfall: 114bl.
National Meteorological Library: 245tl.
Oxford Scientific Films: 25t, 58b; /W Faioley
179c /J Frazier: 159tl & 189tl /B Herrod:
117cr /R Packwood: 121br /V Pared :131bl
/D Simonson: 121cl /K Smith Laboratory &
Scripps Institute of Oceanography: 97br /M
Slater: 121tl /S Stammers: 108c & 111cr /H
Taylor: 97cr /R Toms: 116tr /G Wren: 104cr.
Panos Pictures: 201b & cr, 217cr, 227cr,
234br, 235br & cr. Papilio Photographic: 7t,
29bl, 61c, 13b, 87b, 248t, 250. Planet Earth
Pictures: 6t, 7bl, 8-9, 10c & tl, 11cr, 14tl,
15br, 19c, 25c, 22bl, 23t, 26br, 28tl & br, 33c,
36tr, 37t, 40tl & c, 41tr, c & br, 42tl & br,
43bl, 44tr & br, 48tl, 49tr & br, 52tr & c, 59c,
60b, 64c & b, 65tl, 248bl, 252tr; /Bourseiller
& Durieux: 158br & 183t /C Weston: 143tr
& 154b /I & V Krafft : 131tr & br, 149bl &
171cl /J Corripio: 149cr /J Waters: 133cl /K
Lucas: 159tr & br /R Chesher: 136tr /R
Hessler: 137tl /R Jureit: 161br /WM
Smithey: 160tl. Powerstock Zefa: 112bl,
190br, 212tl, 217bl, 219br & 230bl. Rex
Features: 127, 148tl, 171bl, 172t, 187tr, 239b;
/Pascal Fayolle: 238cr. Science Photo Library:
7cr, 14 t, 48 cl & cr, 59 c, 76tr & cl, 128-129,
131tl, 132tr, 137cl, bl & br, 140tl, 143tl & cl,
144c, 145cl, 166br, 189br, 200tr, 216tr, 217tl,
224tr, 230tr, 234bl, 245tr, br & bl, 249, 251bl,
252bl; /J Autrey: 231l /A Bartel: 242br /M
Bond: 91b /JL Charmet: 170b /L Cook:
175cr /T Craddock: 116br & 221cl /Crown
Copyright, Health & Safety Laboratory: 119br
/E Degginger: 75cl /G Dinijan: 196br /M

Dohrn: 85tl /M Durrance: 213c /G Ewens:
213bl /S Fraser: 96b, 154t, 162t & bl, 163cr,
164tr & 195bl /G Garradd: 167c
/GECO(UK): 184t /F Gohier: 123br /R de
Guglielmo: 158tr /D Hardy: 155l, 166tl /J
Heseltine: 96tl /J Hinsch: 158b /P Jude: 223bl
/K Kent: 225bl /M McKinnon: 117bl / P
Menzel: 171t & 245c /A & H Michler: 122c
& l /NASA: 122br, 167b, 168t & 217br
/Donna & Stephen O'Meara: 243tl /G Olson:
175br: /D Parker: 170t, 182t , 228bl, 231tr &
244br /P Parriainen: 122c & 218tr /A
Pasieka: 156tl /Francoise Sauze: 214tr /S
Stammers: 159c, 168bl /A Sylvester: 81cl /S
Terry: 121br /A C Twomey: 238bl /US
Geological Survey: 168bl, 169t & 171cr /D
Weintraub: 152c. The Skyscan Photolibrary:
13br, 121tr. Frank Spooner Pictures: 149cl,
173br, 175t, 186br & bl, 187tl, br & bl.
Spacecharts: 141cl, 149br, 152bl & br, 168br,
169c & b, 183bl, 204bl, 243tr & 244bl. Still
Pictures: /A Maslennikov: 37t /C Caldicott:
29br /G&M Moss: 17cr. Stockmarket (UK):
216b. Telegraph Colour Library: 146tl.
Topham Picturepoint: 153br, 179bl, 187cr.
Trip/Art: 104br, 194bl, 246tl; /Phototake:
76t. University College London /A.R. Lord:
108l. University of Glasgow /Dr Gribble: 79c.
Tony Waltham Geophotos: 132b, 133bl, 141t
& br, 144b, 145bl, 149t, 153tl, 161t, 163b,
167tl, 183br. Woodmansterne: 155tr.

The publishers would like to thank the
following children, and their parents, for
modelling in this book: Emma Beardmore,
Maria Bloodworth, Tony Borg, Steven Briggs,
Anum Butt, Mitchell Collins, Ashley Cronin,
Joe Davis, Dima Degtyarov, Roxanne Game,
Louise Gannon, Fawwaz Ghany, Hamal
Gohil, Angelica Hambrier, Larrissa
Henderson, Lori Hey, Jackie Ishiekwene,
Daniel Johnson, Sarah Kenna, Jon Leaning,
Louis Loucaides, Catherine McAlpine, Erin
Macarthy, Hanife Manur, Tanya Martin,
Malak Mroue, Griffiths Nipah, Lola Olayinka,
Goke Omolena, Ben Patrick, Daniel Payne,
Anastasia Pryer, Tom Swaine-Jameson, Charlie
Rawlings, Kristi Saxemor, Ini Usoro, Sophie
Viner, Victoria Wallace.

Additional thanks to Caroline Beattie, Mr. and
Mrs. G. R. Evans, Ian Kirkpatrick, West
Meters Ltd and Gregory, Bottley and Lloyd
for their help and the loan of props.

This edition is published by Southwater

Southwater is an imprint of Anness Publishing Ltd
Hermes House, 88–89 Blackfriars Road, London SE1 8HA
tel. 020 7401 2077; fax 020 7633 9499
www.southwaterbooks.com; www.annesspublishing.com

If you like the images in this book and would like to investigate using them for publishing, promotions
or advertising, please visit our website www.practicalpictures.com for more information.

UK agent: The Manning Partnership Ltd;
tel. 01225 478444; fax 01225 478440; sales@manning-partnership.co.uk
UK distributor: Grantham Book Services Ltd;
tel. 01476 541080; fax 01476 541061; orders@gbs.tbs-ltd.co.uk
North American agent/distributor: National Book Network;
tel. 301 459 3366; fax 301 429 5746; www.nbnbooks.com
Australian agent/distributor: Pan Macmillan Australia;
tel. 1300 135 113; fax 1300 135 103; customer.service@macmillan.com.au
New Zealand agent/distributor: David Bateman Ltd;
tel. (09) 415 7664; fax (09) 415 8892

Publisher: Joanna Lorenz
Managing Editor, Children's Books: Gilly Cameron Cooper
Compendium Editor: Jenni Rainford
Additional Design: Michael Morey
Editorial Reader: Jonathan Marshall
Production Controller: Wendy Lawson

ETHICAL TRADING POLICY

Because of our ongoing ecological investment programme, you, as our customer, can have the pleasure and
reassurance of knowing that a tree is being cultivated on your behalf to naturally replace the materials
used to make the book you are holding. For further information about this scheme, go to
www.annesspublishing.com/trees

Previously published as *Amazing Planet Earth*

Bracketed terms are intended for American readers.